ROCKS HEAL!

THE SCIENCE OF ROCK-MEDICINE

By Sela Weidemann

Bowers-Moser

Facilitating change in every direction!

Hardy, Arkansas
www.rock-medicine.com

Published by Bowers-Moser
www.rock-medicine.com

Note to the reader: The information discussed in this book regarding the biochemical properties of the stones is scientifically derived but it has not, as yet, been clinically evaluated. Readers are advised to review this information as formally untested hypothesis. The author of this book does not diagnose, prescribe, and/or dispense medical recommendations for physical maladies without the advice of a physician either directly or indirectly. In the event that you use any of this information in this book for yourself, which is your constitutional right; the author and publisher assume no legal responsibility. This book is not intended to replace medical evaluation and/or prognosis and should not be used to treat a serious ailment without prior consultation with a qualified health care professional.

Weidemann, Sela

Rocks Heal! The Science of Rock-Medicine
ISNB-13: 978-0-692-56539-1

Book Front Cover Design by Joseph Martin
Book Edited by Margaret Schermerhorn
Book Interior Design by Martha Bowers
Back Cover Design by Bowers-Moser
Printed in the United States of America

Dedicated to wellness...

Preface

The intention of this book is to serve as both a layman's use guide, and a comprehensive medical text on the clinical application of crystals and gemstones as medicine. We will show that minerals are precision chemical instruments capable of balancing one's mind, body and spirit to a state of wellbeing, through a system devoid of mysticism, subjectivity, or illusory conventions.

Rock-Medicine differs from other ideological schools of thought regarding crystal and stone influences. Over time the science behind the practice of Rock-Medicine has gained legitimacy. We will show how the laws of chemistry and the laws of quantum and standard physics unite and rule the function of Rock-Medicine. At the foundation of all tangible existence is math and chemistry. All life on the planet, and the planet itself, are made up of varying elements inclusive of Hydrogen, Nitrogen, Carbon and Oxygen. All matter is chemically based and is in motion. This is the fundamental principle of Rock-Medicine.

The business of medicine and medical research today appears to be one of dependency on the existence of illness rather than the eradication of disease. Making the human body into a monopolized marketplace has brought the practice of medicine to its equal and opposite intent. It is time to restore wellness as the right and the responsibility that it is, and not the for-profit industry it has become. Rock-Medicine returns ownership of medicine to all people of all nations. Rock-Medicine does not replace medical practitioners, just the toxic medicines that are currently in use. Rock-Medicine returns us to what nature provides, without synthetics that provide a company with a patented product.

While more and more of us seek alternatives to the toxic pharmaceutical and organic medicines, quantum, or *energy*, methodologies are taking their place. The Rock-Medicine system is that of utilizing the chemical resonance emitted from all sorts of crystals, gems and stones. Rock-Medicine is the only scientifically standardized system of this kind. Minerals can be applied in the same context as pharmaceuticals and naturopathic remedies. By utilizing the chemical vibration and not introducing any residue or additional toxicity, Rock-Medicine is free from harmful side effects. Because you use the chemical vibration of the mineral only, there is no depletion of the source medicine, and this makes for an extremely cost effective program. Rock-Medicine is easily self-administered and shared with others. For example, a single set of the Seven Cleansing Stones can treat the immune systems of an entire village, or nation, forever.

With clear definition, this book is meant to provide comprehensive instruction in the use of the fundamental chemical resonance found in minerals. This text is to be used as a first step to global detoxification and restoration of balance. As we develop a better understanding of how the Spirit and the Mind work, we will continue to expand on the principles of the root causes of physical illness in the human condition. Until then the cleanest, most sustainable and most affordable treatment is Rock-Medicine.

We are three bodies in one. Plotinus (205-275 A.D.) called them "the One, the Intellect, and the Soul". We generally refer to them as the body, mind, and spirit. Illness takes place in the spiritual body first, then the mental, and finally the physical. We start off in this first volume focusing on the tangible physical body. This is the place we human beings are most comfortable. Understandably so, for it is what we can validate with the five physical senses. We relate to our known world by way of what we can see, hear, smell, taste and touch. It is our first priority to heal the physical world in, of, and around us. Appropriately enough, the physical aspects and applications of Rock-Medicine are the easiest to comprehend…so we begin here.

Introduction

Hippocrates (460 – 350B.C.), regarded as the "father of medicine," who authored the Hippocratic Oath, wrote, "By similar things a disease is produced

and through the application of the like, is cured."

Aristotle (384 – 322B.C.) knew the principle as well, and wrote

"Often the simile acts upon the simile."

Rock-Medicine is the application of specific chemicals in rocks and gems to influence areas of the body in which those same chemicals produce, and sometimes impede, vital function. The human being is a complex dance between chemicals and electrical impulses. As you look at medical texts it is amusing to see each particular physiology being emphasized as "vital" to the entire body. It is all vital. Everything is not just connected, but interconnected by a 'nano-fabric' of mass and energy. The only disparity is density and toxicity.

The existence of the physical body is hardly contested. There are multitudes of biochemical observations by the scientific community that evidence the fact that the fittest do survive. This is true whether the organism is pathogenic or not. The life force is *'willful'* energy. This is the survival mechanism that prompts all organisms to adapt to environmental changes.

At the core of wellness is the will, or desire, to obtain wellness. It is instinctive to all terrestrial species to readily engage in combating physical discomfort. We even refer to the pursuit of wellness as the *fight* against disease. All of nature is comprised of survival based, or *solution based*, biochemical activity. There is, however, one significant distinction between Mother Nature's *will* to survive and the use of the word *will* as it applies to human desire. The separation resides in the mind. The human condition is the exclusive result of a physical mass paired with mental cognizance. There is absolutely no debate about the existence of the human mind. It is only the mind of mankind that separates it from natural harmony. The key to healing the human condition, and the world it dominates, is the desire to do so. The human genus' individual and collective wellness, both physical and mental, relies on the perception of wellness. The dichotomy of this dilemma is that the mental perception of true wellness is often skewed by mental disorder. This mental disorder is exacerbated by the increasing toxicity of our physical environment. Human beings are the only life form that engages in cognitive self-destructive behavior. And in doing so, we compromise the progression of personal and social wellbeing.

The flow of wellness is contingent on the absence of toxicity. Toxins disrupt and block the natural pathways of balance. Toxins are generated in many forms, from thoughts, words, and deeds, to the dumping of industrial waste. Toxins are responsible for all that is ill and, at present, all is ill. We have poisoned our planet in its entirety. We cannot kill the planet but we can eliminate human life on it. Man's negative impact can be proven wholly culpable for the extinction of other life forms. We have contaminated all that can be seen with the naked eye and with a microscope. The infiltration of man-made toxins has reached a saturation point. We have tainted all physical matter. We have corrupted the energy space that holds matter together. We are aware of these facts as a people. These practices continue by those who are in social and fiscal control. This is the human condition.

Contamination breeds illness. Unchecked, illness breeds more illness. Now illness is breeding contamination from everything, including its cures. The organic materials used to make medicine, both pharmaceutical and alternative, are the product of polluted air, water and soil. Drugs are, or contain, man made chemical compounds that have injurious effects on the human physiology. Medicine is surrounded by the toxicity of greed and dependency, and it is getting worse. You cannot treat something toxic with

11

something toxic. In other words you cannot treat illness with something that is contaminated. Sustained, long term use of a medicine, as we know it, has become not just useless, but in many cases, harmful. These ancient sacred sciences of stone and crystal already had the answers to the cleanest remedies on Earth.

It is easy for us to see when the life force has gone out of members of the plant or animal kingdom, for they lose their electrical charge. On the other hand, the mineral kingdom does not separate from its electrical charge, as long as one crystal cell remains intact. We are, at the most basic parts of our being, a highly ordered collection of charged minerals. As biological organisms, we cannot survive without minerals. The basic difference between vitamins and minerals is that the body can manufacture vitamins but it cannot make minerals. They must be obtained as elements from nature.

This is why we use the stones and minerals themselves, for their ability to give off a chemically based vibration. It is only by altering them on an atomic level, such as heating, dyeing, irradiating, or pulverizing them, that their elemental electrical charge is damaged or "killed". The whole point in utilizing this medicine is that we can treat ourselves with no residuals, and thus, no side effects. No rock material is ever to be ingested, as Rock-Medicine works via vibration only. Just as a single cell of our body carries the full DNA of our being, so it is with the stones themselves. One small chip or fragment of any stone is just as effective as a huge boulder-sized piece. Each one of the stones' crystalline cells carries the full vibration of that stone's chemical influence. Rocks, crystals, gems and stones are a chemistry set that we can strategically combine for precision clinical applications.

All forms of imbalance equate to toxicity, which in turn manifests as illness. Some stones are used for eradicating toxins while others are used to chemically rebalance ill areas. We have not, as yet, perfected methods of diagnosing and treating precisely the imbalances that take place in the spirit. We have just begun to uncover the means of addressing the imbalances of the mind. This has made our physical body the main focus when searching for healing. The polluting of our waters, air, and soil has rendered the animal and vegetable kingdoms toxic. Again, we cannot treat something toxic with toxic residuals. The plant and animal life from which we derive most of our medicinal products have become virtually useless as a means by which we can obtain health. This catastrophe is echoed in the headlines. Mainstream medicine offers the explanation that our bodies are setting up immunities to common medications. In fact it is the toxicity of the raw materials and additives used to manufacture medicine that our immune systems are attempting to combat. Pollution and over processing, combined with a mindset of maintenance rather than healing, have derailed the concept of true health. Rock-Medicine is the cleanest form of medicine.

We have a window of opportunity before global increases in radiation levels compromise our last living kingdom, the minerals.

Chemical electric vibration allows minerals to impart a medicinal resonant tone used for healing. This chemically based vibration travels throughout our body as an electrical resonance carried by the water content in our body. It is important to drink adequate amounts of water to keep the body hydrated so the pH level, electrical conductivity, is at an optimum. All forms of toxicity result in physical malady, whether it is altering the quality of source material by excessive processing or the polluting of its host environment. All loss of wellness is wholly due to toxic blockage somewhere in our individual and collective being. Rock-Medicine comes at a time when mankind's efforts are turning towards the health of our planet. This planet has provided for the survival of life here. Our conscientious use of the available resources guarantees the quality of that life. Protecting our world from the toxicity produced by selfishness and greed insures longevity of life and of its quality to be well.

Rocks, stones, crystals, and gems are the ideal chemical tools at this time in the planet's evolution. Formed eons ago, they are free from man-made pollution. With them we can detoxify ourselves, and the planet. Nothing living can tolerate for long an environment that is toxic to it. We are learning that lesson as we track the changes in the health of certain life forms and the extinction of others. If we expect to keep company with cosmic consciousness, we have to clean up the toxic condition of our world's spirit, mind and body. Rock-Medicine is help for the life forces of this third dimension. It is not a magical influence on other dimensions, parallel existences, or individual whim. The animal, vegetable and mineral kingdoms are all relative to each other, as are the mind, body and soul of an individual. All life energy is commonly shared by the total amount of matter and space in our dimension. Given this oneness, the health of the one affects the wellness of all. Just as the chain is only as strong as its weakest link, so can all of creation only be doing as well as we as individuals are.

The density of the mineral kingdom made it the perfect place to store medicinal chemicals frozen at a time before pollution. Through Rock-Medicine we can treat the common cold or AIDS. We can use this science for healing or prevention.

The only requirement for these minerals to be effective is that they be used according to all instruction and safeguards. The methods are simple and few. No faith or affirmation is necessary. We all have the ability to manifest our own wellness but are not in a clear enough space to exercise it. If we were using more than a meager percentage of our brain's capacity we would not even be having this conversation. Rock-Medicine will facilitate healing ourselves to a point of truly utilizing our individual and collective powers to their unlimited potential. Since the minerals' chemical vibrations are a constant, we gain consistency in effective treatment.

Rock-Medicine enables us to treat one another with or without knowledge and/or permission. Denial of illness and refusal of treatment are illnesses in and of themselves. We must take individual responsibility for the whole if we are to advance.

Crystals, rocks, gemstones, fossils, metals, and even crystallized saps such as amber, make up the vast array of precision chemical vibration instruments used in Rock-Medicine. As the time-space continuum moves us through cosmic change, we see antiquation as we advance. This is also true of the applications of the stones. Ancient standards are no longer applicable and have evolved into a different correlation to the planet, atmosphere and inhabitants. The Code of Hammurabi is an excellent record and example of the evolution of justice and wellness as we compare it to what we hold to be standards today. We have an obligation to discover the validity of all inspired sciences. Modern advances in data management and communication make such clinical research testing available at little to no cost. The number of people on the planet seeking alternative medical practices is now the majority. This has heightened the responsibility of both scientists and medical practitioners to enter into a new cooperation with alternative medical means. Leading physicists and medical researchers are currently documenting new scientific discoveries that validate the basic concepts that make up the foundation of Rock-Medicine. Many ancient practices such as acupuncture, tai chi, yoga, meditation and creative imaging are emerging as modern means to the promotion of wellness. We must begin now to incorporate newly awakened sciences into our mainstream methodology if we are to have a chance at the survival of humankind and the sustainability of the garden in which we live.

A Brief History of Medicine

The earliest known written text regarding both the practice of law and that of medicine is the origin of the "eye for an eye" reference in the Code of Hammurabi written in 1780 B.C. The portions pertaining to medicine, translated thus far, speak more to the compensation and conduct of medical practitioners. Little to no information has been recovered from the Code as to the actual practice of medicine itself. The extensive tablets containing the Code are found to have a list of medicines called the 'materia medica' and include plants as well as over 100 minerals.

The oldest recognized medical text is the Edwin Smith Papyrus dated 1600 B.C. It represents advanced knowledge of the mechanics of medicine. It is the oldest copy of a trauma surgery textbook. It gives exquisite detail of the human anatomy. We remain a species adept at the manipulation of that same anatomy. Fifty years later, the Ebers Papyrus was written, although it's believed to be copied from texts as old as 3200 B.C. The papyrus is the second oldest medical text ever found. It is the largest of the ancient medical texts and incorporates the knowledge of anatomy with holistic practicality. It represents the approach of body, mind and spirit with their respective sciences. They may be regarded as remedies, potions and incantations. These two Egyptian texts come from a time and place still represented today by unfathomable human accomplishment.

Traditional Chinese medicine is derived from the same philosophy. It is the classical belief that the life and activity of individual human beings have an intimate relationship with the environment on all levels. In his *Bencao Tujing* ('Illustrated Pharmacopoeia') the scholar-official Su Song (1020–1101) categorized herbs and minerals according to their pharmaceutical uses. Although well accepted in the mainstream of medical care throughout East Asia, it is still considered an alternative medical system in much of the modern Western world.

Paracelsus was the sixteenth century physician who led the way in using chemicals and minerals as medicine. In 1605, at 16 years old, he entered Basel University's medical school. Other than his medical doctorate, his fields of expertise included alchemy and astrology. Paracelsus believed that there is a willful energy that connects all that is, and manipulates harmony, and that this energy can be identified.

In some ways, the beliefs and practices of eighteenth century German physician Franz Mesmer look back to an earlier period of magical medicine two hundred years before his own time, to the era of Paracelsus. He studied medicine in 1759 at the University of Vienna. Aside from being a medical doctor, Mesmer is credited as being one of the fathers of modern day hypnosis and psychotherapy. He believed that subatomic ether operates in individuals. When it flows naturally, the result is a normal healthy condition throughout creation. Should the natural flow of this ether be impeded in any way, sickness is the result. Sickness affects the body, mind and spirit collectively and individually.

Recent identification has been made of something scientists call *dark matter.* Glenn Starkman at Case Western Reserve, and colleagues Tom Zlosnik and Pedro Ferreira of the University of Oxford are now reincarnating the ether, dark matter, in a new form. Their aim is to solve the puzzle of the mysterious substance. They anticipate it will explain why galaxies seem to contain much more mass than can be accounted for by visible matter. They pose that ether is a field, rather than a substance, that pervades space and time. "If you removed everything else in the universe, the ether would still be there," says Zlosnik. According to Starkman this ether field isn't to do with light, but rather is something that boosts the gravitational pull of stars and galaxies, making them seem heavier.

It does so by increasing the flexibility of space and time. Within the subatomic structures in the human body, this is the unseen life force.

Once only theorized, after repeated calculations on the universal mass and gravitational forces, it proves credible. This validates the practices of some of the earliest biorhythmic energy studies, including those of Anton Mesmer. Mesmer's 1770 basic theory is remarkable in light of new discoveries in the realm of latter 20th century physics. While many still dismiss his therapeutic successes as only applicable to hysterical or imagined illness, Mesmer could demonstrate cases cured by his treatment that had previously failed all conventional approaches. Even his detractors conceded this. Some of his patients went on to lead quite functional lives, where before they were deemed hopeless invalids. His claims of dramatic therapeutic success were supported by glowing testimonials, and in some cases, from socially prominent individuals. However, mainstream medical practitioners, professional societies, and politicians including Benjamin Franklin, rejected Mesmer and his treatments, and ultimately moved to eliminate Mesmer's practice and that of his peers.

The Southeastern Pennsylvania region is a leader in bioscience innovation. The nation's greatest concentration of leading medical/bioscience research institutions is located there.

A 1681 partial repayment of a debt, from Charles II of England to William Penn, created a charter for what would become the Pennsylvania Colony. Despite the royal charter, Penn paid the local Lenape tribe for the land so as to be on good terms with the Native Americans and ensure peace for his colony. As a devout Quaker, Penn had experienced persecution. He developed his colony to be a place where anyone could worship freely. This extreme tolerance led to significantly healthier relationships with local Native tribes than most other colonies experienced. It fostered rapid growth, resulting in Philadelphia being America's most important city. Penn named the city Philadelphia, which is Greek for 'brotherly love'. William Penn hoped that Philadelphia would become a community modeled more like an English rural town instead of a city. He laid out roads on a deliberate plan to keep houses and businesses spread far apart. He believed this would stimulate sustainability by allowing them to be surrounded by gardens and orchards. However, the city's inhabitants didn't follow Penn's plans and crowded by the Delaware River, and then subdivided and resold their lots. The city soon established itself as an important trading center, poor at first, but with tolerable living conditions by the 1750s.

It is at this time that Benjamin Franklin was instrumental in having the development of western medicine be limited to only the anatomical knowledge of leading physicians at European medical schools, based on mummification practices and mummified corpses acquired in Egypt. He was a participant in filtering out holistic study and practice, as evidenced by his campaigns against the works of Franz Mesmer and other proponents of energetic medicine. In less than one hundred years, thousands of years of traditional and esthetic, or holistic, medical practices were done away with by a handful of overzealous doctors and merchants in eighteenth century Philadelphia.

Understandably driven by the desire to hold true to a medical science devoid of fraud, these early influential American pundits ended up "throwing the baby out with the bath water." It has taken nearly two centuries for western medicine to return to the science of nature instead of limiting the nature of science. In today's world the influences on wellness include such studies as the effect of attitude, stress, music, color, humor and even prayer.

In 1821, Philadelphia was the largest city in the United States at 137,000 strong. Originally the country's capital, 1821 Philadelphia was the site of the first American law school, hospital, and college of medicine. It was then that druggists and apothecaries organized themselves and created The Philadelphia College of Pharmacy. We had the world's first school of pharmacy connected to a university and the beginning of the pharmaceutical industry. The school was an idea pioneered by John Morgan in 1755 when he founded the Medical School of the College of Philadelphia. His conviction at the time was that there should be a separation of the practice of pharmacy and the practice of medicine. Prior to this, Physicians made and sold their own remedies. He said that a patient "ought to know what it is they really pay for their medicine and what for medical advice and attendance." He noted the importance of physicians not limiting their practices to selling and using only those remedies produced by the individual doctors themselves. What we call a conflict of interest.

Even though Franklin failed to understand the relevance and existence of the esthetic sciences, he still understood the underlying principle that we are all one. He understood that wellness should be a right in this country. The first hospital at the time, Pennsylvania Hospital, dispensed medicine freely to those who could not afford it. This followed the second primary tenet of Morgan's belief that the dignity in medicine was that all citizens would receive necessary treatment and/or medications regardless of their social or financial status. This spirit of humanity was reflected in the fact that the hospital received a steady flow of illnesses, which did not change a great deal from 1776 to 1948. These patients were all poor because the Philadelphia Hospital's rule, in Benjamin Franklin's own handwriting, mandated service to the "sick poor, and only if there is room, for those who can pay."

Philadelphia is home to America's first hospital, Pennsylvania Hospital, founded in 1751. The first hospital for children, The Children's Hospital of Philadelphia, was founded in 1855. The first independent medical research facility in the United States, The Wistar Institute, was established in 1892. The first medical school and teaching hospital was the University of Pennsylvania School of Medicine, founded in 1874. The first cancer hospital was the Hospital of the Fox Chase Cancer Center, founded in 1904. The first college of pharmacy, University of the Sciences, was founded in 1821. The first private psychiatric hospital, Friends Hospital, was founded in 1813. All of these located in the City of Brotherly Love, which is, ironically, the home of this author's birth.

The region continues to be the primary support base for the nation's Pharmaceutical Industry, an industry that today has a reputation for its wealth-based agenda and undue influence on Washington itself. Attached to what we know as healthcare today, we have not only the environmental poisons of air, water and soil pollution, and over processing, but also the toxicity of greed. There is too little motivation for true healing. We have become a society of medical maintenance as opposed to treatment and prevention. In retrospect, it is clear that traditional physicians have had little to offer their patients therapeutically that had any real possibility of benefit. Doctors are taught in schools that are supported by Pharmacia. They do not know what they do not know. They are for the most part taught how to treat symptoms or to cut. Devastating side effects attributed to toxic pharmaceuticals are the legacy of modern medicine.

Cutting-edge western medical pursuits engage in the adaptation of "attitude" being integral to wellness. Our attitude is as much a part of our physiology as any other process i.e.; auto immune responses, where all functions of the body are produced and regulated. We can deduce it is subtle electric brain chemistry that directs our attitude or "being". This is where wellness and its opposite, illness, reside. We must treat the spirit of an individual in order to treat the body. It is now accepted without question that attitude has everything to do with one's wellness. Natural mineral therapy as applied in Rock-Medicine, allows for the least radical, least invasive impact on an already fragile and compromised body's chemical constitution.

Allopathic, or mainstream western medicine, searches for treatments for symptoms as opposed to "cures" for diseases. This approach ignores blockage in the natural operating system (the nervous system), and its regulatory systems (the immune system), that contributed to the original illness. Viewing such as "sick" or "broken" instead of blocked asserts in practice that there is something that must be attacked and killed in the system, instead of working to address what is blocking the system. Only in surgery do we actually engage in a repair tactic.

Herbal and other organic medicines stop short of eradicating illness because they still are bulk based, organic composition medicine filled with the toxins of the air, water and soil that produced them. This brings with it residue which elicits additional toxicity resulting in "negative side effects".

We've seen the awesome ability in some individuals to beat chronic and fatal pathogen replication and "heal themselves" by what appears to be sheer will. This is demonstrated again and again in clinical trials where half or more of recipients in double blind studies "heal", even when given a placebo, or even when half of a group responds well to a drug and others given the same drug do not.

Rock-Medicine differs from all previously perused methodologies. It deals with the "fields" of energy, making it unified physics, as opposed to merely the parameters that traditional physics presents. Minerals are precision instruments capable of balancing to a state of well being one's mind, body, and spirit. Their application is non-invasive, non-toxic, non-residual and cannot be rejected or mutated by the patient's systems. If illness is present, then so is toxicity. How can a chemical overload ever seek to bring "wellness" about if pathogens are highly adaptive cellular structures that proliferate in toxicity?

Impoverished areas of the global population are more susceptible to pathogen invasion by virtue of the depletion of the human spirit. Deprivation of food, clothing and shelter, and the absence of peace and prosperity by oppression, contributes to epidemics. These stresses of spirit have a direct impact on "dis-ease". To date we have had no applied science for the human spirit. Standards developed across three decades of clinical application show we can use natural minerals for their chemical-electric properties in a targeted capacity.

Rock-Medicine works by way of electromagnetic vibration based upon chemical resonance. We are matter in motion; this is grade school science class information. All that exists is minute particles moving at rapid speeds. Though we cannot pass our hand through a tabletop, we know that it actually consists of individual atoms in constant motion. Those atoms have individual elements as well. Subatomic particles are particles smaller than an atom. In 1940, the five subatomic particles known to science were limited to the primary building blocks of atoms being protons, neutrons and electrons. The two other particles, neutrinos and positrons were discovered outside Earth's atmosphere. Twenty years of scientific discovery, including atom smashers, nuclear fission, and fusion, saw the list grow to nearly one hundred.

In 1964, American physicist Murray Gell-Mann and Swiss physicist George Zweig established that the five originally listed elemental subatomic particles were indeed correct. They represented the fundamental particles that cannot be broken down into any simpler particle. The subsequently listed particles were discovered existing at a higher than normally observable energy. They were particles similar to the proton, neutron, electron, and the other two curious subatomic particles, but having one property (such as an electric charge) opposite them. It was established that certain particles emit various types of forces. An example of this is the photon. It is a particle that contains no mass yet transmits electromagnetic energy from one place to another.

The three life kingdoms on the planet give off electricity when they are alive. This force is known in Quantum Physics as a charm quark. It is one of the few remaining physical science frontiers. These elusive building blocks of living matter are defined as the nuclear force that holds matter together. Having to do with the little known physics of gravity, these forces top the list of scientific theory for the cosmic action known as the Big Bang and the origin of the life force. It is this chemically charged electrical force that is exchanged between the minerals and the biochemical bodies they treat.

The mineral kingdom is being recognized as a viable means of healing, and history substantiates this kingdom's work record. The very names of individual stones sometimes were given them because of their healing virtues. We see countless examples of ancient tombs safeguarding the remains of the dead and the amulets and beads used to treat them while alive. Both historical reference and prophetic prediction of the clinical use of the mineral kingdom are well documented. The source of Rock-Medicine's specifics is the same as the knowledge vault credited by Edgar Cayce and Carl Jung for their information. The akashic record has educated many of the great teachers, some we even call masters of enlightenment. It is credible theory backed by inspiration and science combined.

Contamination and Cleansing of Stones

The tools used in Rock-Medicine require minimum maintenance. The rules that apply, however, must be strictly observed. You begin by obtaining minerals. Only unadulterated material is acceptable. Stones must not be heated, irradiated, pressurized, dyed, or stabilized with other substances such as plastic. Rough uncut specimens work as well as polished and/or carved. Size does not matter nor does quality. Whether you purchase, dig, or are gifted stones and gems, always make sure each stone is correctly identified. Sometimes suppliers substitute a stone of similar appearance for the actual mineral you seek. Verify that your source is reliable and the specimen is unmodified.

Next, you want to be sure that the piece of rock be not mixed with another separate mineral. One example of this would be Azurite and Malachite. These two minerals often grow together. You can see the presence of each quite clearly. The blue Azurite is a single instrument intended for a specific purpose. The same goes for the green Malachite. When combined they form a third chemical influence. If you acquire material that is piggybacked with another mineral, they need to be separated. Breaking crystals off a matrix or chipping a portion of the desired material off another mineral will not impede or diminish their effectiveness.

When the terms "clear" or "clean" are used in reference to the minerals, it relates to energetic contamination. Obviously, if there is a foreign substance on the stones, such as oil, dirt, food etc. you want to wipe them off or wash them in water. Using mild soap is okay, if necessary. A few minerals, such as Halite and Sulfur for example, are salts or chalklike. Materials such as these are water-soluble and are cleaned with a soft brush or cloth; use no water and NO soap.

While minerals are too dense to be contaminated by environmental pollution, they are affected by contact with animal and plant life. They have a mass, and so, a subatomic spin. Because everything is matter in motion, all mass gives off a vibration. A chemically specific electrical energy is commingled when one life form touches another. In all three living kingdoms, toxins are exchanged by contact. In the case of gems and stones, they immediately mesh with the ether of toxins present in any life form that touches them. Stones and "spreads" used in Rock-Medicine must be protected from all contact with persons, animals and insects. Even the slightest skin contact will trigger toxic energy exchange. Rock-

Medicine tools need to be regarded as sterile instruments. When you handle them to make an essence, or set up a distance spread, you should wear gloves, or you may use non-metallic items for transferring stones, such as wooden or plastic spoons, chopsticks, or plastic tweezers. When minerals are held for a medicinal application they need to be cleared before being used again. If stones are touched in any way they must be cleared before use.

Just as the cosmos is in constant motion following set paths of revolution, so is the molecular universe of the minerals. Their crystalline structure is spinning like a miniature galaxy. It takes a single minute for our blood to completely circulate one time throughout our body. Every two to four weeks, new epidermis replaces old skin. Stones cleanse themselves by completing a full revolution, or cycle, of their atomic substance in a matter of three hours.

Stones may be cleared all together. They may be cleared in any non-conducting non-chemical container. A glass bowl or cotton bag makes no difference. Whether they touch each other or not doesn't matter. To clear the stones it is not necessary to do anything but leave them alone. No sunlight, moonlight or salt cleansing makes any difference. In fact, using salt crystals directly on minerals, whether sea salt or table salt, is detrimental. Salt has a caustic effect on minerals and will result in microscopic "pitting" of the crystals. This damages them.

Stones will clear best if left alone in locations not attached to electrical current. On a table or bookshelf is fine. It is more important to keep them at a distance from metal and/or electrical influences including televisions, computers, stereos and appliances. Minerals must remain untouched by any other organic life form during the three-hour cleansing process. In three hours time all minerals convert themselves to a clean vibration of their own pure chemical distinctiveness. It is this resonance that is utilized for healing. The minerals' vibration broadcasts a chemical charge that can be captured in water or transmitted from place to place by quartz crystal, but only clean mineral vibration can elicit balance and wellbeing.

Ill energy can likewise be transmitted. Energy is real and can be felt, whether ill or well. For example, it is what we can feel in a room where two individuals feel love towards each other ...or animosity. It is what we refer to when we say the tension in the air could be "cut with a knife". So, just as thoughts give off a tangible, if not measurable, vibration that affects other life forms much more so do minerals give off chemical vibrations that affect the life forms they contact. The body is an elaborate collection of billions of cells, all cooperatively multitasking in unison with each other and all other matter. It is important to always work with clean, contaminant-free tools. Understanding the stones and how they relate to body function and form is what learning the language of Rock-Medicine means.

Setting Stones

Certain stones in the combination may be dangerous to handle, such as Cinnabar with its mercury toxicity. The stones are, after all, a chemistry set, and some chemicals are hazardous when they come into skin contact for prolonged periods of time, or as in the case of the rock salt Halite, is of a structure that may "melt" and leach into the skin. Some minerals can leach chemical particulates into the skin of the hand. Some of these are cinnabar, halite, sulfur, and would include any specimens that are shedding residue. In a clinical venue, it is important that all applications of Rock-Medicine be executed with the utmost vigilance. If particulates leach into a patient's system, there will be problems with residual effects. In these cases, this author advises using a 'set' crystal for the hand-held method. This is unless and until one possesses the knowledge that a given stone may be safely held with regard to its stability and content.

When a piece of clear Quartz crystal is in contact with other stones, it takes on, or *records*, the combined vibrations and stores them for a limited period of time. (See "Clear Quartz Crystal") A Quartz crystal can be "set" to record and store the chemical resonance, and then may be hand-held in place of the stones themselves.

When you have your combination of stones chosen, put them aside for three hours to clear with a single piece of Quartz crystal in with them. All stones and the crystal must touch one or another stone in the grouping. The crystal will "set" to the stones resonance and clear of toxins at the same time. For each hand-held application you will hold the crystal, instead of the stones, for just short of twenty minutes. When you put down the crystal after use, it will be returned to the same grouping of stones to both clear and reset. You may at times use a group of stones where one needs to be a set crystal. To simplify, use a crystal set with all stones in the combination at once. (See Setting Stones) This regimen is repeated four times a day at 3-4 hour intervals for a period of weeks to months, as determined by the patient's need and their individual case.

The Four Methods of Application

There are four methods of application for Rock-Medicine:

1) *Hand-held*

2) *Essence*

3) *Focus Direct*

4) *Blanket Spread.*

From the first to the fourth, they are listed in order of greater to lesser ability as far as bringing the most expeditious healing. Hand-holding the stones is the strongest, fastest way to measurable results. Using a rendered essence is the next most expeditious. A focus direct application works quite a bit slower, as does the blanket spread. Holding stones will produce results within minutes to hours; essence, from hours to days. A focus direct take weeks to months, and the blanket spread takes months to years to elicit measurable results.

Deciding which method to use is determined by many factors. Taken into consideration first will be how convenient it is for the patient. Although hand-held is the strongest use of Rock-Medicine, it is a moot point if the patient is either unwilling or unable to engage in its faithful use. As hand-holding the stones require up to 20 minutes four times a day, time management becomes a factor. It is better to use the essence if there is better guarantee of consistent compliance. These two methods may be combined. Hand-holding and essences are applied four times a day at 3-4 hour intervals. While you may only use one combination at a single application you may switch up methods for ease. For example, hand-holding stones for twenty minutes in the morning for your first application may work well for you, but could be challenging for additional applications throughout the day. You could find it more convenient to hand hold stones once, take essence at three hours and then six hours later, and then hold stones for the fourth and final application at the end of your day.

In the case of someone who cannot participate in their treatment due to being mentally or physically incapacitated, and they have no full time caregiver, a focus direct is their best help. The focus direct method treats a room with a constant resonance. It may be directed to a specific place in the room as well. Blanket spreads enable us to treat hundreds and even thousands of miles in radius. As with a focus direct, the blanket spread requires no conscious participation on the part of those to be influenced. A large enough spread can treat the whole planet at once.

Determining the length of time to continue a given treatment will depend on the age of the patient, combined with the severity of the condition and the longevity of the condition. The age of the patient represents the saturation level of environmental pollutants. These toxins are responsible for the failure of the immune system. The longer a person is exposed to the air, water and soil pollution, the longer it takes to detoxify and repair their immune system. For this reason alone, a child will respond more quickly to Rock-Medicine. The time frame for treatment with regard to severity and longevity is going to differ greatly. Consider the comparison of recovery expectations from the flu vs. Parkinson's for example. The flu can take up to 48 hours to fully clear where Parkinson's can take up to 6 months.

When doing a hand-held or essence application one must remove **any** jewelry, stones, or metals that they have on. Once jewelry is removed, hold your stones or take your essence. Jewelry may be put back on immediately following the application and removed again for subsequent applications. This is to prevent any interference from other metals and stones that could change the chemical equation being used for the application. Be mindful of contact with electrical sources like remote controls or cell phones while holding stones or taking essence.

Methods may be combined for individual or multiple treatments that run concurrently. For treating multiple conditions at one time, the applications must be done separate from each other. If treating diabetes at the same time as treating the eyesight, single combinations are applied one after another. Stones or essence for diabetes could be followed by an application of stones or essence for vision.

Since a single hand-held application takes twenty minutes, four times a day. One condition treated by this method would involve a total of one hour and twenty minutes of hand-holding per day. Adding a second regimen for a separate condition using the hand-held method increases this by another hour and twenty minutes. Few people have that kind of time to spare in their busy schedules. You can see where having the option to place a couple of drops of essence under the tongue four times a day saves a lot of time. Treating multiple conditions with essences is much more practical. Certainly, individuals who have the time to adhere to the schedule of a hand-held application should do it. Usually this will apply to those who are too ill to engage in normal life activities such as a job or social events. Their illness *is* their day.

Rock-Medicine combinations form a sentence. Each stone added to that sentence will direct the healing vibration more and more specifically. The stones communicate with each other by chemical resonance. Rock-Medicine is a precision technology. Consequently, if you were to combine the stone for the eyes with the stone for the spine, neither area could be addressed. This would be a conflictive combination and thereby, ineffective. The stones would be conflicted as to what area they were being directed to go. They cannot go to both the eyes and the spine at the same time. This confusion of focus will negate the effectiveness of any type of application.

Almost every combination starts with the Seven Cleansers. Jade, Amber, Smokey Quartz, Pyrite, Clay, Cobalt, and Hematite or Carnelian depending on age, are the chemical balance of the immune systems. No matter what the imbalance, complaint, or malady, usually healing with Rock-Medicine begins with the decontamination of the energy pathways in the immune systems by use of just the Seven Cleansers. This is especially true when more than one medical condition is present. Unless a patient is critical, and loss of life is eminent, we nearly always recommend starting with the Seven Cleansers. In an emergency, we do use specific stones for a given malady or area in the body to stave off function collapse. Likewise for chronic conditions, we will give a formula for the specific condition at hand. Then, when the patient is stable, we begin the Seven Cleanser application. A minimum course of the Seven Cleanser balancing of the immune system would be three weeks. Some patients will want to continue for up to 6 weeks depending on age and condition. As stated, once the immune system becomes operational again, many symptoms will disappear within days. After 3-6 weeks of Seven Cleansers the patient is re-evaluated and specific remaining conditions may be identified and treated with more precise formulas of stones.

Please note that when you first begin to use Rock-Medicine you will experience an initial rapid detoxification on all levels. It is common to have several hours of flu-like symptoms a day or two into treatment, as the body eliminates toxins through whichever orifice is most convenient to the locale of the affected body part(s). This could be in the form of vomit, diarrhea, or increased urination. It may be secretions such as mucus from nose or lungs and even fluid or higher volume wax from the ears.

The more hydrated we are, the better Rock-Medicine will work. Drinking half your body weight in ounces of water per day represents the minimum standard requirement for survival. If you weigh one hundred pounds, your daily water intake should be 50 ounces. It is one of the simplest things we can do for ourselves to restore and help sustain wellness.

Rock-Medicine is not contraindicated for use with any other medicines. Where they are used, it is important to get medication levels periodically checked by a physician, especially in the first three weeks, as the body heals and no longer requires them. Whether the medical regimen is herbal or pharmaceutical, it is important that one not be over-medicating. Rock-Medicine stabilizes the chemical energies in our bodies, and that balance will be jeopardized if taking invasive medications unnecessarily.

Finally, it is *always* necessary to get a reliable diagnosis from qualified medical professionals. Only by verification of disease can the correct combination of stones be determined. This is medicine and must be used as such. Rock-Medicine precisely targets illness. Different stones will be used for an infection than would be used for cancer, for example. Often more than one condition is present. They may be related or not. Each condition will require a separate combination of stones. We do not use guessing, intuiting or reflexology to ascertain which stones are called for. The problem with relying on these methods is that if the patient has multiple areas of imbalance, the stones indicated for use may not be all that are needed, or are contraindicated for combination with each other.

Hand-Held Method

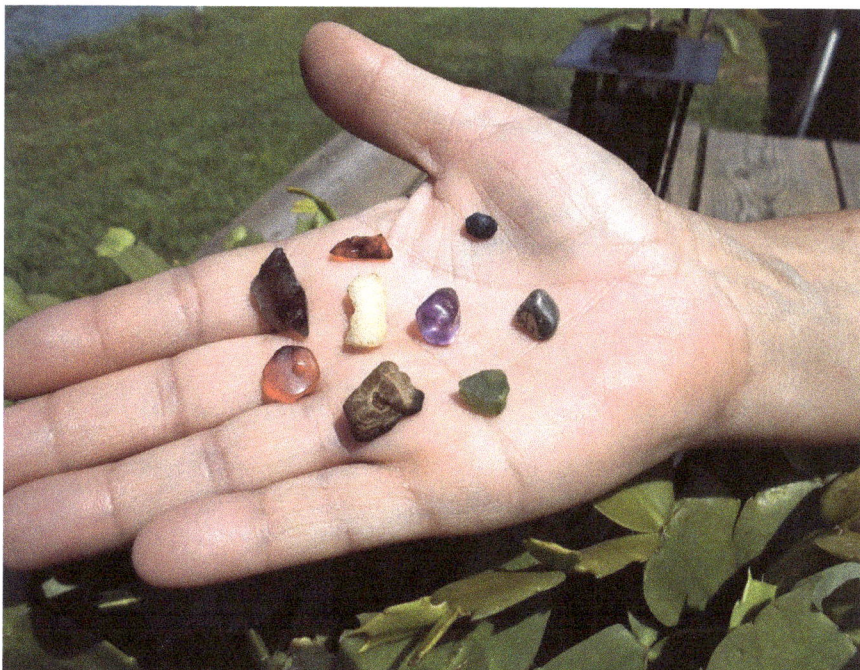

As previously stated, Rock-Medicine is a mineral-based chemical energy system. It produces energy forces that travel throughout our body as a chemical electric charge. The water content in our bodies provides the delivery system. Contact with the stones for physical medicine is always made at the extremities. Following the same guidelines as acupuncture does, women's energy flows from her right to her left; men's energy runs from his left to his right. The chromosomal pattern designating male and female is the basis to determine which hand to use. This is the yin-yang flow of our body's electrical system of meridians. With hand-holding, all stones combined for the specific imbalance being addressed are to be held together in the proper hand. For men, the proper holding hand is the left. For women it's the right. This is consistent with the direction of a body's energy flow.

The hand-held method is the strongest most expeditious means of administering Rock-Medicine because of the level of contact. It is direct. The stones themselves will have the strongest resonance. The stones are held for up to, but not exceeding, twenty minutes. This is consistent with the common atomic motion cycle. If the time exceeds twenty minutes, by even a second, you run the risk of over saturating, or over mingling, the stones' vibrations with the toxic vibrations they contact in you. The result will be the voiding of that application, but it will not make you any sicker than you were. Eighteen minutes is a minimum for a hand-held application. After the holding time is up, set the stones aside to clear for at least three hours. This is repeated four times a day at 3-4 hour intervals.

Once all jewelry is removed, be sure that you have arranged to have your holding hand occupied for twenty minutes. Optimally you will not be touching anything electronic for the duration of the application. Many things in our environment put out EMR (electro-magnetic radiation). Some EMR output is limited to close contact with small electronic items such as phones, games or remote controls. Other systems such as computers and televisions can emit

EMR several feet. Since the stones are performing at a subatomic level on subtle electrical energies we limit their exposure to electronics while clearing and when in use.

Choose carefully your stones for combination. Be sure that the equation or *sentence* that the stone influences create is concise and complete. (See "Working With Combinations") There is no requirement for limited activity other than what is necessary to consistently hold the stones in the proper hand for a period of twenty minutes. Do not touch other living organisms while doing an application. Both plants and animals are able to exchange toxic energies.

The effect of Rock-Medicine is akin to the way that prescription antibiotics work. When the rock vibration encounters an ailing area, the stone(s) chemical message reorganizes the biochemical energies of the area being treated and boosts it up towards its point of well balance. After three to four hours, the wellness level begins dropping back down towards the ill, or low, resonance level. Then you come in with the next application. Over time, these repetitive intervals of chemical adjustment act like a parent teaching a child to ride their two-wheeler. The balance point is learned, or re-learned, by the body, and eventually holds itself at a constant. The energetic alignment is intended to ultimately be self-sustaining, eliminating the need for further treatments. This is the difference between a healing and a maintenance program.

A single malady can require as many as twelve stones in a single combination. Be mindful of this when obtaining stones. Make sure they are not so large as to be cumbersome, and not too small to keep track of, as sometimes the chemical equation contains a large number of stones.

Essence Method

A Rock-Medicine "essence" is water that contains the electrical and chemical charge of one or more minerals. Water is a good conductor of electrical currents. The chemical vibration coming off the crystals and gems is in the form of an electrical charge. We can capture that same vibration in a receptacle of specially prepared water for use in the form of an essence.

It is the water in our bodies that conducts and distributes the stones' nuclear influence. That is why pH is one of the first things to be looked at if you are experiencing imbalance in your body in any way, shape or form. Our body's pH level regulates our electrical system and intracellular activity, as well as the way our bodies utilize enzymes, minerals, and vitamins. Likewise the essence solution used in Rock-Medicine has to be of the correct pH in order for the chemical resonance to be integrated by the body.

Tap water, spring water and purified water all contain trace minerals, which could alter the prescribed stone formula, and so are not reliable water sources for essence making. In order to render an essence that is purely the vibration of the intended stones, we use distilled water. However, distilled water is free from all mineral particulates and so has an electrical conductivity rating of zero and a pH of 7.0. This means that the distilled water is unable to conduct any electrical current on any level.

The human constitution is fragile and resonates at a lower frequency when ill. Medicine should not exacerbate a condition or assault other body systems. In order for distilled water to conduct stone resonance we need to introduce negative ions for the charge to travel on. Adding non-iodized salt to the distilled water does this. We choose kosher salt over sea salt because sea salt has over eighty naturally occurring minerals in it that could alter the chemical formula of your chosen stone combination. Processed table salt often contains anti-clumping agents and other chemicals. Pure salt is sodium chloride and is not contraindicated for use in an essence. Sodium chloride, being naturally present in the human biology, will not change the information in your stone equation. Its presence will do no harm to the set crystal, the essence, or the patient. The salt content will not cause any adverse reactions if other medications are present in the patient's system. Be sure you do not touch the salt crystals with bare skin or metal utensils so as to observe rules regarding contamination. In order to achieve salinity equal to human blood, the ratio of salt to water for one gallon of distilled water would be one teaspoon of kosher salt. The exact level of salinity necessary for conductivity for the purpose of essences is unverified at this time. I am going with the blood match of salinity for now, absent conclusive information to the contrary. When fully dissolved, this amount of salt introduces necessary particulates into the distilled water that facilitate the capture and distribution of the stones' vibration.

BPA-free plastic or non-leaded glass is the best receptacle for the rendering and storage of essence. Metal containers cannot be used. The stones' electric energy would contact the metal sides, once introduced to the salinized water, and then travel around its exterior instead of being captured.

Clear Quartz crystal, as mentioned before, can be used to "record" the chemical vibration of stones. When rendering an essence we use a set crystal to insure that no mineral residue gets into the water. Essence must be of a purity that guarantees it can be safely ingested. The absence of residual particulates from the stones guarantees this quality. Some minerals are water-soluble. Some leach toxic material when submerged in water. Some form rust. All these are potential health hazards.

Having chosen the stones you want to combine for use, place the stones together with a Quartz crystal. These may all be piled on a non-metallic surface or plate or in any non-metallic cup, bag or bowl. Not all stones need to touch the crystal themselves, but all stones must touch one another and/or the crystal in some way. Where you place the stones to set during this process is your choice. In or out of a receptacle makes no difference. These will remain untouched for three hours.

The Quartz crystal will record the stones' vibration in three hours. The crystal may then be removed using gloves or plastic utensil to avoid contact contamination by your skin. The crystal may then be dropped into your container of salinized distilled water. In three hours you will have a full strength essence. Think of what happens when you drop a pebble into a pond. The ripples migrate outwards until they hit the shore. Then they are absorbed into the land buffer. When you drop a charged crystal into a container of water the vibrations ripple outward, hit the sides and rebound, intersecting each other and infusing the water with the vibrations. This is why we do not use a metal container. These vibrations remain in the water until use. No preservative or suspender is needed beyond the already present dissolved salt.

After three hours, the vibration is permanently captured by, and held in the water. The crystal may then be removed from the water if desired. The essence retains its full strength. This essence may not be diluted for use. However, if the crystal is left in the essence, as the water level goes down with use, you may simply refill it with more of the distilled water solution and the crystal will continue to charge it. Essence can be made in any volume.

If a patient is required to take nothing by mouth, per physician's orders, essence may be applied by drops to the wrists or forehead, although this is not as effective as ingesting it. Essence works extremely well for animals by way of drops poured into a clean, non-metallic saucer or dish.

The standard application of essence is a few drops held briefly under the tongue before swallowing. The capillaries under the tongue are very close to the surface and absorb essence quickly into the bloodstream. Hand-held as well as essence applications follow the same schedule of four times a day at 3-4 hour intervals. More will not do more. There is no possibility of overdosing with Rock-Medicine, but drinking distilled or salt water in large quantities is dangerous. The same rules apply here as for all applications regarding duration of use. The age of the patient combined with the severity and longevity of an illness determines how much time is required for treatment.

Once you have made a container of essence, the threat of contamination is still an important consideration. Essence itself must not be touched. Even taking a sip from a bottle of essence and having it touch your lips will contaminate the essence with backwash. If an eye dropper is used it is vital that the dropper not come into contact with the interior of the mouth or saliva will contaminate the essence once the dropper is returned to the bottle. Squirt and dropper bottles of various types are all viable options. One could pour a small amount, ¼ tsp., of essence into a glass and drink it, holding the essence under the tongue briefly before swallowing.

Essence can be stored anywhere and does not require refrigeration. If you make a large quantity that will not be used for a while, put it in the refrigerator to keep it from light and heat.

Focus Direct

In order to work a focus direct application, the first thing required is a Quartz crystal with one or more perfect points, or terminations, on it. This means crystal point(s) with no chips, cracks or damage to them of any kind. Over the years the question arises as to whether or not a damaged point can be re-ground to a pristine condition. Until we have absolute verification one way or another by credible scientific evaluation we are not advising the use of re-ground points for this application.

When a stone or stones are placed against the base or sides of a quartz crystal with a perfect point their vibration travels up to and "out" the point in a laser-beam like action. (See photo) This beam can be directed at an individual and constitutes an application of Rock-Medicine. This beam will not travel through walls or glass or turn corners. If anyone steps between the beam and its intended goal, that person will then be its affected target.

If a cluster of Quartz crystal is used, a beam will be emitted from each and every perfect point on the cluster. The stone(s) may be placed anywhere on the cluster. The vibrations travel in the same direction as the termination(s) points. The distance the influence goes is based upon the total weight of the quartz used. If a one pound single point is used, a single beam goes out one hundred feet. If a one pound cluster is used, a one hundred foot beam is emitted from every pristine termination on the cluster, no matter how large or small. The sends the beam(s) a distance directly proportional to the weight of the piece of quartz crystal used. That distance is one hundred feet per pound. A two pound piece sends chemical resonance two hundred feet, a half pound piece sends it fifty feet, and so on. Rarely is a room even close to fifty feet in any direction, so you can see it does not take much to set a focus direct up. (See "Quartz Crystal")

This makes an excellent tool to use in any home, office, classroom or workplace. A focus direct can be aimed at a bed, an easy chair, or a therapist's table. The focus direct can be set for any single malady, stone or combination of shield stones. In an office or classroom it may contain Granite for kindness,

Petrified Wood is for stress, and Chrysocolla is for academic achievement. It could be set for diabetes or Alzheimer's. The focus direct takes much longer to elicit a state of wellness than the hand-held or essence method. Weeks, or even months may be necessary. Still, it is an extremely utilitarian option given the degree of need worldwide.

Both the Quartz crystal and the stones must be cleared before assembling a focus direct. The combination of stones must be a singular combination. As with other methods, only one specific combination at a time can be used with a single Quartz point or cluster. Touching cleared stones contaminates them. Touching a focus direct contaminates it, so use safe handling guidelines. Since the focus direct will not work if enclosed or covered, it needs to be disassembled and cleaned and cleared periodically to ensure it is not contaminated by human, animal or insect contact.

Blanket Spread

A blanket spread is also generated through clear Quartz crystal. The crystal or crystals used need not have good points or terminations. It is just the bulk weight of quartz itself that is being used. There is no limit to the size or number of quartz pieces you can pile together as the generator. A blanket spread will go a distance equal to one hundred miles per pound of quartz crystal. Planet Earth has a diameter of 7,918 miles. A blanket spread with one pound of quartz will treat a radius of one hundred miles. A spread containing 80 pounds would treat our entire planet, her inhabitants and eight thousand miles up through the atmosphere into space. A blanket spread will penetrate all manner of material. A blanket spread may be placed inside any type of a case to protect it from dust particulates and human, animal, or insect contact.

Blanket spreads can take from 2-4 years to elicit results. A community, city, state or planet can be treated for diabetes. Or, it may have its air pollution relieved.

We can have multiples up at the same time, however. For example, two spreads can be set up side by side or on different parts of the world that both contain 80 pounds of quartz crystal. One could be addressing malaria, and one to address leukemia. They will work independent of each other and pose no conflict with each other or any other Rock-Medicine use.

Global response can be expected to be measurable in 3-5 years. It would be prudent to have this type of effort be cooperative, since two global spreads for one illness or purpose would be redundant and a waste of resources.

Place cleared bulk crystal on any non-metallic surface. One piece or many different pieces piled all together can be used. The total weight will be your distance. Stones chosen must be for single specific combinations. Clear the stones (See Contamination &

Cleansing). Place your combination of stones for use on or against the quartz. All stones must touch crystal. Quartz must be placed so there is contact between them all.

Lastly place a cleared Lapis Lazuli strand around the spread at its base. It does not matter if the strand touches the spread or not, as long as it surrounds the base completely. (See photo) Lapis is the "on" switch, so to speak. Removal of the Lapis strand shuts the spread down. I have found that turning an aquarium over on the blanket spread to protect it from dust particulates, insects, and other contaminating contacts enables me to still see the spread and monitor it as well as enjoy its beauty.

A well-protected blanket spread does not have to ever be disassembled for clearing or cleaning. Once again, all precautions are to be followed so that a blanket spread does not send contaminated energy out.

The Power of Clear Quartz Crystal

Quartz crystal is used in Rock-Medicine to *record, amplify,* and *transmit* the chemical electric resonance given off by other stones and minerals.

When we say clear quartz Crystal we are referring to quartz that is absent color. There are varying degrees of clarity from opaque with iron deposits to brilliantly transparent, like glass. Clarity is not relevant to the use of quartz crystal in Rock-Medicine. It is the total absence of color that is important. Quartz crystal, as with all other

Rock-Medicine tools, is always used in a form free from other mineral matrix.

Clear quartz crystal is one of the most abundant materials on the surface and crust of the planet. It is the most stable form of silicon dioxide. Silicon dioxide is found in two places in nature. The first is quartz and the second is in the cell walls of *diatoms*. Diatoms are living breathing algae that emerged at or around the Jurassic Period. Diatom fossils are a reference we currently use to study environmental conditions and water quality, both past and present.

The first published studies of ionic, or electrical, conductivity relating to quartz crystal was conducted in 1952. An **ion** is an atom or molecule where the total number of electrons is not equal to the total number of protons, giving it a net positive or negative electrical charge. Quartz crystal has a monovalent, or positive only, ion.

The quartz molecules' parallel arrangement of *trigonal* a-quartz to the Z-axis causes an inherent electrical charge. Trigonal refers to a molecular geometry of one atom in the center of three other atoms forming an equidistant triangle. The Z-axis is depth, creating a three-dimensional plane complete with ether and thus, the electric charge. (The crystal structure of quartz is a simple atomic model of the third dimension.) Positive ions move along a crystal's shaft towards its point. They are off in search of a negative charge, traveling pathways created by the quartz's atomic structure. We can detect and observe the life force of a quartz crystal. It is this now measurable spark of life...the marriage of gravity and electromagnetism...that delivers the first tangible evidence of the elusive unified field.

Quartz is known for its uses relative to resonant conductivity. Over the past century quartz has been found in timepieces, radio apparatus, and thermal wave regulation. It has been processed for a wide variety of electrical components. Quartz crystal is a vibration amplification and storage system.

When animals, vegetables or minerals touch quartz it immediately begins to record and amplify all biochemical fields. Quartz, like all minerals, will engage in energy exchange from its own kind first. Quartz seeks the presence of other minerals and interacts with them first.

This means that if you are wearing an emerald ring the contact you make with quartz will amplify the chemical energy field of the emerald and metals in the ring. If you have no other minerals in contact with you for the quartz to detect, it will amplify the energy field of the next life force...which is you. Quartz will amplify whatever is going on in your entire body. It makes no difference whether that condition is well or ill, the quartz will amplify it. If you are happy, healthy, well, it will amplify it. If you are sick, sad, angry, depressed...it will amplify it. It is harmful to you and to others when we engage in the spontaneous amplification and exchange of *ill or poor* energy. We already do this daily. We expose each other to communicable disease all the time. Sometimes we transmit virus or bacteria, and other times infect others with our anger or depression. Quartz crystal makes these exchanges more invasive by intensifying them.

Just like all other minerals, quartz crystal clears itself of foreign vibration after three hours of being left alone, away from EMR and organic toxins. (See Contamination and Cleansing) Once quartz has been cleared, it holds no toxic vibration from other organic life forms or matter. A cleared crystal holds no message of its own and so, has the potential to be used in a variety ways.

Although all quartz is able to record stone resonance, the manner in which it transmits it can depend on its size and/or structure. From chips and chunks to points and clusters, quartz is available in many forms. Some Rock-Medicine applications call for perfect points, or terminations, others for no points at all. In a focus direct and blanket spread the weight of the quartz will be taken into consideration. Not so for the hand-held or essence methods.

The electromagnetic energy of quartz always moves in the same direction. The positive ionic charge moves from base to point. It follows the same path as the crystals' growth, up and out.

When holding a set crystal or using a set crystal to render an essence the size does not matter. Neither method requires the quartz to have a defined termination. The termination refers to a quartz crystal's point. Sometimes they are just a single point. Sometimes we have a cluster which is multiple points protruding from a single base. Natural quartz points occur in every imaginable size. From the minutest of *druzy* to the world's massive giants, weighing tons each. Quartz is fragile in that it can fracture, chip or break, like glass. Crystal points are easily damaged. A perfect point is a natural termination point that has no damage to it at all. There is only one application in which having a pristine point(s) is integral. This is the focus direct effect (See "Focus Direct").

The charge that runs along a crystal will actually carry a *hosted* energy out and beyond the physical structure of the crystal, if there is any perfect point present. This hosted energy could be a mineral vibration or the energetic imprint of another life form.

When we touch a piece of quartz an imperceptible electrical charge snaps between our skin and the crystal, because the crystal emits a life force in the form of an electrical charge just as we do.

Our energy, well or ill, can flow from the point of a perfect quartz termination just like a stone vibration, and go a distance equal to a hundred feet per pound (See "Focus Direct"). If we are touching a cluster our energy goes out every one of the intact natural points. If anyone is within those energy beams' line of sight and range, they will be affected. Therefore, we must exercise caution when touching clear crystal points and aiming them at another person.

A perfect point is not required for hand-held, essence rendering, or blanket spreads. Single points or clusters may be used for any of the four methods but the integrity of any points present is only relevant to the focus direct.

Quartz crystal can record stones' energy field. The set quartz can then be used in a hand-held method. A 'set' quartz may also be used to infuse the energy field of water with the imprint of the energy field of a group of stones. After a 'set' crystal has been in a salinized distilled water solution for three hours, the essence is ready and is full strength. For use and storage, it does not matter if the crystal is removed or left in the container of essence.

If you put one or more pieces of cleared (see "Contamination & Cleansing") quartz into a glass bowl, you can pour essence into the bowl to cover them. Three hours later, the quartz piece(s) will have recorded the vibration from the water. A set crystal will hold the imprint of a stone combination for three hours out of water. If not put into salinized distilled water within that time, it will clear of the original stone formula.

Each piece of quartz may then be taken out of the bowl of essence. You could then place each quartz piece into a different glass receptacle. Pour salinized distilled water into the receptacle to cover the crystal piece you removed from the bowl. Three hours later each of those containers is full strength essence with the crystal in it. Again, the essence is full strength and the crystal can stay in or be removed without affecting the essence.

In review, a crystal can be 'set' to record other stones. That crystal may then be used to charge a saline-water solution. The resulting essence can be used to set more crystals which can then be used to make more essence. A 'set' crystal will charge water with any stone combination, and any essence may be used to charge a crystal. (See "Essence")

Once you have set a crystal to a stone combination, you have endless distribution potential for only the cost of salt, distilled water and quartz. Quartz crystal allows us to dispense endless amounts of Rock-Medicine. Stones and essence can be safely mailed or flown with. There is no danger to it from the security scanning devices. The only danger from contamination is by direct human contact with the water or the crystal. Always wear gloves or use non-conducting implements when transferring quartz to and from water.

Start with the Seven Cleansers

As mentioned before, virtually all Rock-Medicine treatments will begin with just the Seven Cleansers. Jade removes blockage. Amber gives the body its memory of wellness back. Pyrite and smoky quartz purify the air and water in our bodies, respectively. Depending on age, either Hematite or Carnelian purifies the blood. Cobalt eliminates radiation build-up. Clay regulates the combined immune systems. This grouping of stones represents the immune system as a whole. The immune systems of the body, mind and soul are all "jump started" with Rock-Medicine. When illness is self-perpetuating, the only course of action is to *re-polarize* the body's energy. This has to happen in the *nanoparticles*.

The immune system is a multiple-system structure that identifies and kills tumor cells and pathogens. With an immune system in good working order no antigen or pathogen that attacks is able to invade. Virtually all illness is connected to the failure of the immune systems. Almost every treatment combination will include the Seven Cleansers as its base. With Amethyst, the eight stones together represent the cellular body only. Toxins compromise the immune systems of all current life forms wherever they occur. Our autoimmune functions are as adversely affected by stress as by carcinogens.

Once our initial protection is compromised, disease invades very quickly. Often multiple conditions are present. The Seven Cleansers alone will begin the detoxification of the whole being it is applied to.

Our immune systems are the primary functions in our bodies equipped to maintain wellness. These natural defenses are integral to the process of healing. The combined systems establish, maintain and defend all normal balances. When the immune system itself is compromised, there is no chance to restore or sustain a healthy body.

Listed here are the seven general cleansing stones and their properties. As one reviews the individual and combined influences, it is clear how basic the function of the immune system truly is. These stones are used all together in a single application. They form the foundation upon which all other physical Rock-Medicine remedies are built. All of the Seven Cleansers can be used effectively in any of the four methods of application.

Jade

We begin with Jade. Nephrite Jade is calcium magnesium. Calcium and magnesium are the components that initiate detoxification within the human body. Nobel Prize winners in Medicine, Dr.Bert Sackman M.D. and Biochemist Edwin Neher have discovered and proven through many scientific studies that macro minerals such as calcium and magnesium must be converted into ions before they are absorbed within the human body. The presentation of Rock-Medicine as electric chemical ether is naturally that ionic form.

Jade is for releasing blockage that has disrupted and prevented the flow of wellness. These blockages are the result of toxic build-up. The use of this stone goes back to ancient Egypt, Central America and China. It was known as the "side stone" or "hip stone" because of its healing influence on the kidneys.

The kidneys are responsible for homeostasis. Homeostasis is the ability or tendency of an organism or cell to maintain internal equilibrium by adjusting its physiological processes. The kidneys filter toxins out of the blood. The elimination of residual waste from our body is at the core of preventing blockage and illness. As all things evolve, so to do the stones and their association to our physiology. Jade's influence is no longer limited to influence on the kidneys, but seeks and breaks down blockages throughout our physical, mental and spiritual bodies.

As any illness denotes a blockage in the energy flow, Rock-Medicine uses Nephrite Jade for every toxic condition. It is considered the "grandfather stone" of healing, as eliminating blockage is the first step necessary to restore health. Jade influence is what causes the body to release toxins from the body in fluids, often creating flu-like symptoms in the first day or two of initial use.

As with Agate, Jasper and other stones, the color of Nephrite Jade used and the quality, are not factors in its use or application. Jade is jade no matter what, as long as it is natural untreated Nephrite Jade. The two jades, jadeite and nephrite, can be somewhat similar in appearance. Not as flashy as its cousin jadeite, nephrite is the true jade of ancient China. The form of Jade used in Rock-Medicine is the Nephrite.

Jade can be applied topically to the temples for fast relief from a headache. Another variation of use is anytime a noticeable degree of heat or fever is present. The patient then applies the jade directly to the area for a period of twenty minutes. This topical application works well in many instances for immediate relief from pain and/or swelling. This type of remedy is just for temporary relief and is not intended to replace a schedule of treatment for healing with jade as included in the Seven Cleansers combination.

Amber

Amber is the stone that connects our bodies to their memory of wellness as contained in our DNA. Amber is fossilized tree resin. Being fossilized conifer resin, it is crystallized abscisic acid. Abscisic acid is known to activate nucleic acid synthesis. **Nucleic Acids** consist of two distinct, but closely related chemical forms: deoxyribonucleic acid (DNA) and ribonucleic acid (RNA). The main functions of these biomolecules include the storage of all heritable information of all organisms on earth, and the conversion of this information into proteins.

78

1. The DNA double helix uncoils and the two strands separate.

2. RNA Polymerase attaches to promoter regions of the DNA strand

3. RNA polymerase binds free RNA nucleotides to create mRNA that are a copied template of the corresponding DNA strand.

4. The mRNA separates from the DNA.

5. The DNA strands reforms into a double helix.

Because the nitrogenous bases that compose DNA can only pair with complementary bases, any two linked strands of DNA are complementary. This ensures that the old base sequence is conserved and uncompromised. It is the RNA messaging from the DNA to the state of being that is corrected, using Amber as the memory of wellness.

Amber is one of the oldest crystal formations on the planet. Once the Jade has removed the blockage, Amber steps in to carry the memory to the area afflicted, of what that area's proper state of wellness is. Amber uses your own DNA to lock on to a previously well state. Because DNA holds the patterns of every one of your ancestors, if you are born with an imbalance, Amber will search back as many generations as necessary to find a healthy strand to lock on to and use as a guide to map the path that equals balance. In other words, if you were born with diabetes, as in Type 1, Amber looks in your DNA strand for a parent, grandparent, or

Start with the Seven Cleaners

great grandparent whose genetic makeup was not predisposed to the condition. It then pulls the proper genetic configuration from there.

Memory of wellness applies as equally to the physical broken leg as it does to the mental and spiritual attributes that caused the break in the first place. It is this memory of balance that is responsible for the stones ability to restore blood pressure to a well state or blood sugar to its proper level. The chemical vibrations of the stones restore that sense of balance by Amber's restoration of the body's memory of wellness.

Human DNA consists of chromosomal DNA and mitochondrial DNA.

The chromosomal DNA is located in the cell nucleus. It is responsible for storing an individuals' hereditary genome. Mitochondrial DNA is responsible for the conversion of food into a form of energy able to be utilized by cells. When the vibration of Amber is introduced to the human body, it immediately links to the DNA to find an example of healthy chromosomal DNA and function from the mitochondrial DNA.

The other stones you include in the formula determine the specific location and nature of the malady. When used with just the Seven Cleansers, the Amber uses the memory of wellness stored in the

80

DNA to find the chemical vibration representing a healthy immune system. The energetic representation of the body with a healthy immune system is then shared with any ailing areas of the immune system. Amber instructs any given area as to how its form, function, or both, are supposed to be.

Hematite or Carnelian

Carnelian   Hematite

Beyond water, the most essential fluid in our physical body is blood. Blood is the delivery system to our cells for all necessary substances such as nutrients and oxygen, and it carries away waste toxins for elimination. Our blood supplies oxygen to tissues. It supplies nutrients such as fatty acids, amino acids and glucose. It transports waste away, including carbon dioxide. Blood flowing through our circulatory system is the physical conduit to move things through the body – whether oxygen, antibodies or hormones.

Coagulation, or clotting, is also a function performed in the blood. Blood is the physical communicator for the transport of hormones. It signals the presence of tissue damage. Blood regulates the body's pH as well as core temperature. Our blood assists us with motor function. A vital fluid indeed! Hematite is iron oxide and our blood is essentially iron oxide.

Hematite is the blood purifier for the younger generation, or, those under the age of 55. Carnelian, a translucent, semiprecious variety of the silica mineral chalcedony, owes its red to reddish brown color to the incorporation of small amounts of iron oxide.

Carnelian is the blood purifier for those fifty-five and older. Our biological systems change drastically at this "mid-life" point. Hormones and metabolism change. Our ratio of yellow to red marrow increases from the day we are born. Red marrow is where blood cells are produced. (See "Iolite") Cells and cell processes evolve with the aging process, making parts of our body very different than in our youth. Toxins are also different, in both their chemistry and density, as the generations pass.

The name Hematite is derived from the Greek word for blood. Carnelian is named for its color, which resembles that of a type of cherry known as the Cornel. When putting combinations together, one or the other stone will be chosen for blood purification, based upon the age of the patient. Hematite is for those under 55 years old. Carnelian is for those 55 and older. This rule is not affected by any other variables. Women who have experienced early menopause or someone afflicted with an aging disease like progeria would still abide by their birth date. Do not use both hematite and carnelian together. Don't send a mixed message regarding the age of the patient, which is easily found out. (For animals it is actual calendar years not calculated in animal years)

These two stones are a good example of why we use a 'set' crystal (See Setting Stones) to render an essence of stone resonance. The Carnelian may be safely put into water because it is silica based and the iron content is structurally contained. Not so with hematite which would begin to react in the water and produce rust. It is always important to follow safety guidelines! It is safer and more reliable to always use a 'set' crystal.

Pyrite

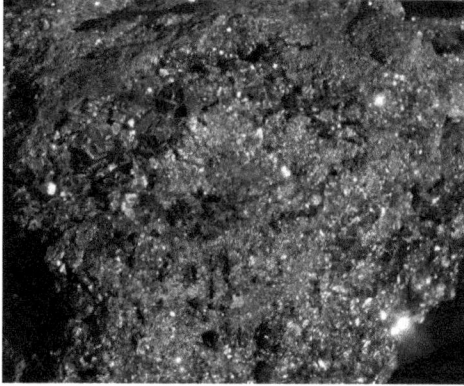

Pyrite is for oxygen purification in the form of iron sulfide and directly impacts ferritin protein. Compound iron is central to respiration, photosynthesis, and DNA synthesis, through its interactive relationship with oxygen. A ferritin is an intracellular protein that is found in virtually all-living organisms. Ferritin proteins store iron and release it in a controlled fashion. It acts as a buffer against iron deficiency and iron overload. Ferritin proteins manage iron *and* oxygen. Ferritin protein function is central to molecular nutrition, cellular metabolism, at the crossroads of iron and oxygen in our body's biochemistry.

Pyrite will treat impurities in the oxygen content in our bodies as well as in our environment. It does not just clean the air in our orifices, but purifies the oxygen content in our cells. The human body is approximately two-thirds oxygen, rivaling water as the majority of our mass. Oxygen is obtained through the lungs and absorbed by the lungs' vascular system. From there it is elaborately processed throughout the body's cells. Pyrite enables your body to absorb the oxygen it needs without interference from other molecules. For example, nitrates can mix with the blood taking the place of oxygen carriers. When this happens to a child, we call it blue baby syndrome.

All of Earth's living organisms, with the exception of some bacteria, thrive on oxygen. The importance of oxygen to life is vast and complex. Researchers have established that many individuals are functioning at as much as a 50% reduction of necessary levels. Oxygen is both the method and mechanism by which our bodies convert matter to energy. Low oxygen levels are among the first signs of illness.

Pyrite's resonance interfaces with oxygen molecules in members of the animal and vegetable kingdom as well as in soil and the atmosphere. Pyrite's purifying influence elicits both the production of and quality of oxygen. Our current levels of soil, water and air pollution have compromised the balance of the whole planet and all her inhabitants. Many herbal and flower remedies are used as medicine, and they themselves are currently too toxic to be effective. The same is true of all our resource material for food. Oxygen molecules are everywhere.

Smokey Quartz

Smokey Quartz is for the purification of water. It contains tiny clusters of aluminum atoms whose vibration extract pure hydrogen from water. In reference to the notion that "what kills a thing cures it", a negative environmental effect of aluminum is that its ions can react with phosphates, which causes phosphates to be less available to water organisms. Those phosphates, or your pH level, are essential to good health. Likewise the pH balance regulates proper electrolyte levels. This stone's influence flushes toxins out of H2O in our bodies. H2O is what facilitates the elimination of toxins as part of the normal body function. Although we can survive weeks or so without food, we can only go mere days without water.

To function properly, the human body requires anywhere from one to seven liters per day. Why the wide range? A wide range of variables, such as weight, exertion, temperature, etc., prevents standard intake values for any given individual. As a result, and for the same reasons, water content in the human body ranges from about 58% – 90% as babies, to 55%-78% in adulthood, making up the bulk of our mass. Aside from being elemental to many of our various cell structures, body fluids make H2O vital to a multitude of processes and functions. Synovial fluid serves to lubricate joints. Other fluids act like shock absorbers such as surrounds the brain and eyes. Water is important to hydration. Hydration is how our body regulates its temperature through processes such as sweat and evaporation. Water is part of virtually all our physiological components.

Impurities ingested from the water we drink and the water we irrigate our crops with, are processed by our liver. Elevated saturation levels of toxicity in our H2O are higher than can be disposed of. This creates residual buildup of toxicity in the liver. Aside from fuel management and detoxification, the liver regulates the distribution of water throughout the body systems and organs. Using Smoky Quartz as part of the Seven Cleansers will flush toxins from our body's water content at all levels.

Cobalt

Cobalt is used for radiation toxicity and the thyroid gland that it impacts. Pure cobalt is highly radioactive and unstable. For the purpose of Rock-Medicine we may use any one of a number of rocks that incorporate a safe form of cobalt and like compounds. These may include Heterogenite, Linnaeite, Covellite, Zag meteorite, Cobaltine, or Cobaltite. Because it is the paramagnetic inertia we are utilizing, as opposed to the chemical resonance, we have this wide range of stones to choose from. Most commonly used in the decades of our work has been Covellite.

Intact thyroid peroxidase showed a homogeneous high-spin

electroparametic signal with axial symmetry. Many cobalt composites and the stone Covellite are metallic conductors with weak Pauli-Para magnetism. With Pauli-Para magnetism, only an electron near the Fermi Surface can change its spin to align with the magnetic field. (This is as opposed to the Curie Paramagnetic spin)

The thyroid gland is the most radiation sensitive part of the body. In the animal kingdom, radiation is filtered by the thyroid gland. It is also the seat of our inspiration and our anxiety. The TPO gene provides instructions for making an enzyme called thyroid peroxidase. This enzyme plays a central role in the function of the thyroid gland.

The thyroid controls our body's metabolism, converting all that we take in into the fuel that drives us. Some fuels are more toxic than others. Our whole planet and its environment and atmosphere are all toxic with radiation. The planet has natural levels of radiation, which we have driven dangerously high. Industrial waste, technological devices, EMRs (electro-magnetic radiation), weapons testing, and mining are major contributing factors to the rising levels.

Radiation is the only influence that can compromise the life forces of the mineral kingdom, putting at risk our last source of clean medicine. Planetary levels of radiation are on the rise and it is this influence that can put the mineral kingdom itself in danger. Although air, water, and oil borne pollutants cannot impact the minerals, radiation does.

The function of the thyroid gland is to take iodine and convert it to the unique hormones that regulate the conversion of oxygen and calories into energy. Ionizing radiation is taken in through everything we ingest. Being part of the endocrine system, this in turn affects our mood, growth, hormones, metabolism, and sexual and reproductive functions. The thyroid collects radiation toxicity. Radiobiologists have long believed that ionizing radiation, like gamma rays, kills cells by shattering DNA. As we always say in Rock-Medicine 'like' heals 'like'.

Given as a common treatment for radiation toxicity, iodine collects in the thyroid. It takes up space that radioactive iodine would occupy. Without room to stay in the thyroid, the radioactive iodine is sent out of the thyroid and excreted with urine.

The ability to treat the thyroid gland is the most sophisticated tool we have. We say this because the most advanced form of human-made toxicity is radiation.

Cobalt is for the thyroid gland itself. As the gland structure is not gender specific you will use cobalt in any sentence that is addressing the gland itself. Cobalt is essential to the repair and restoration of the immune systems. Cobalt may also be used, in a focus direct for example, to treat food, herbs and the stones themselves, for radiation toxicity. An application of at least 24 hours would be a minimum requirement.

In terms of the emotional body, the hormones produced in the thyroid influence one's imagination. Our imagination is what enables us to "see potential" and advance towards it on all levels. Whether we imagine the best or the worst is reliant on the health of the thyroid gland. A positive attitude alone is able to lift one out of illness or even prevent it.

Clay

Clay's aluminosilicates impact the regulation of the immune systems. Many different cell types make up our immune systems. They are all contained in our body...the clay vessel. The immune systems protect us from the successful invasion of antibodies. Utilizing Clay's subtle energies for brain chemical cascade to the spinal cord helps heals the auto immune system and its counterparts to prevent invasion as well as replication of pathogens.

Specific combinations of minerals adjust our brain chemistry to release toxic blockages that compromises the immune system. The immune system is then able to prevent pathogen invasion.

Illness begins in our spirit, compromising the mindset and ultimately the physical body. The human being is a delicate balance of all three, and it is therefore necessary to treat all three at the same time. The planet and her other living kingdoms have their own immune systems. Clay is only applicable to the animal kingdom. And to humans, it is hope.

When we speak of using Clay as an instrument of Rock-Medicine, it is always in reference to a piece of pottery that is in excess of 200 years old. Material needs to be obtained from eras predating the Industrial Age. It is in that period that soil, water, and airborne pollutants had not yet contaminated all surface and subsurface clay on the earth. Since the natural density factor of clay is so slight, it has absorbed all of the toxic elements over modern times. The man-made toxins over the past two centuries are at a dangerous saturation level. If you use a piece of clay pottery or tile that is over 200 years old, you are insured against getting an impure vibration. The process of making pottery is to mix clay with water and then to sun or fire-bake it. This increases its density factor. It is no longer susceptible to residual pollutants. Old clay shards are not as difficult to obtain as you might think. Old pottery, tiles, and other clay household, ritual and decorative items are actually in abundance.

Combining the Stones

A wide variety of stones, when added to the Seven Cleansers, complete a formula or an equation for specific physical body combinations. These will give you the accuracy in targeting an area of the body and its systems that are treated with Rock-Medicine.

Some stones give us an indication of *where* in the body the imbalance is located, such as the eyes, specified by the stone Malachite. Some stones tell us *what* is affected at the location. Cinnabar for example is indicated for the location of the heart or the heart itself; alone it says treat the heart. By adding Garnet you would be pinpointing a vein or artery in the heart.

Halite specifies muscle. Malachite with Halite says, "eye muscle". Coral indicates a calcification. Malachite with Coral says "calcification on the eye" or cataract. These combinations can be viewed as creating an intelligible sentence with stones.

Example: **7 Cleansers + Amethyst + Lepidolite + Rhodonite**
This grouping of stones is equal to:
The Immune System + Cell Division + Brain + Neurological

This combination indicates a treatment target of a neurological disorder of the central nervous system, which is located in the brain. If the stone Lepidolite were to be eliminated, it would change the combination to one for any nerve damage in the body in the external nervous system, or peripheral nervous system, not being located in the brain. Adding other stones can redirect the formula such as:

7 Cleansers + Amethyst + Rhodonite + Emerald

This would be used for nerve damage in the back.

ALL applications of Rock-Medicine will include the Seven Cleansers for the immune systems. *ALL physical applications will include Amethyst for cell division.*

The Stones Themselves

Abalone

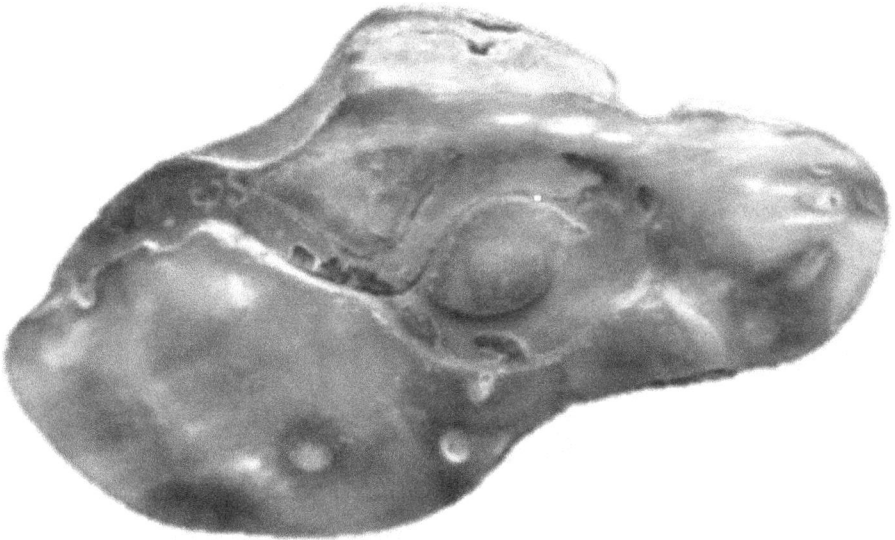

Abalone is Calcium carbonate and is for the hippocampus. The hippocampus is the part of the brain that is involved in memory forming, organizing, and storing. It is this area that we find the obvious application as it relates to Alzheimer's and dementia. A complete combination for the hippocampus would include the seven cleansers, amethyst and abalone. Once used in Native American puberty ceremonies for young girls this stone can also help with the literal organizing of ones home space.

Amethyst

As a silicon dioxide with iron, Amethyst triggers and accelerates cell division. The word "amethyst" comes from the Greeks, meaning literally not toxic. Amethyst is used in all applications where cell division is mandated. We are referring specifically to cell *mitosis* as opposed to cell *meiosis*, which is embryonic cell division during sexual reproduction. Cell mitosis is nuclear division plus cytoplasm division, which produces two identical daughter cells. This is replication vs. reproduction.

Cells divide whether they are well or ill. Cells divide when tumors grow or infections spread, where skin is burned or bones broken.

This is the reason Amethyst should never be used or worn alone. Amethyst makes no determination as to whether the cells are dividing in a healthy or mutating fashion. It just makes cells divide faster. If cell disruption is present in the form of illness or injury, the disruption will be worse.

Anytime an application is strictly a physical one, Amethyst is included. Its one and only message is "cells divide". Wearing Amethyst alone would make the condition of any ailment worse. It will cause the occurrence of any injury to be worse when it happens. If you burn yourself it will be more severe. If you cut yourself it will be deeper. Amethyst receives its direction from the stones you add to it. When treating a healing broken leg, for example, combining Amethyst with Coral, (for bone cells) sends the message to only promote healthy bone cell division. Amethyst will be withheld from combinations intended to treat the mental and spiritual bodies. Not only is it not applicable, it confuses the total stone formula as its presence denotes a physical treatment exclusively.

Aquamarine

Aquamarine indicates the mouth. It is beryllium aluminum cyclosilicate with iron. The inclusion of this stone in a combination sends the mineral resonance to the mouth and mouth only. This is 'where' the imbalance is, the mouth. You may also have to specify 'what' the afflicted part of the mouth is. This is where the art of combinations comes in.

If you add Coral to the Aquamarine you are saying that any calcium based part of the mouth is the targeted area. This would be either teeth or jaw bone. Combined with Rhodocrosite, indicating

100

epidermis would designate gums or skin. Halite would indicate muscle. In the mouth it could be the tongue or jaw muscle. You could not combine Coral with Halite for example, as this would be making a statement regarding the tooth muscle. This is not a logical equation with respect to a treatment target area.

As has been stated, Rock-Medicine does not replace medical practitioners. Damaged areas of the mouth may be treated with Rock-Medicine to an extent. Dentists still need to fill cavities and manual manipulations will continue to be done by them. Afflictions including infections and abscesses will be treated with Rock-Medicine. Other areas aforementioned are subject to repair and restoration using Rock-Medicine, i.e.: muscle, bone, adhesion and injury.

Rock-Medicine can be applied to address inflammation and pain until qualified dental assistance can be obtained. These instruments are not meant to replace conventional medical practices as a whole. Rather, they work in conjunction with them for healing, as opposed to maintenance.

Azurite

Azurite is a basic copper carbonate mineral. Simple copper carbonate ($CuCO_3$) is not known to exist in nature. Azurite has the formula $Cu_3(CO_3)_2(OH)_2$, with the copper(II) cations linking it to two different anions, carbonate and hydroxide. Acetylcholine increases smooth muscle relaxation. Smooth muscle is responsible for the contractility of smooth muscle as can be found in hollow organs, such as blood vessels, the gastrointestinal tract, the bladder, or the uterus. There are many examples. Combining Azurite with yellow Calcite would indicate precisely the smooth muscle tissue of the bladder, for example. Combining it with green Calcite pinpoints the smooth muscle tissue of the gastrointestinal tract...with Pearl, the uterus...etc. Knowing the stones is the key to understanding all the nuances!

Azurite combined with Garnet (blood vessel) and Lepidolite (brain) will facilitate smooth muscle relaxation of vessels to increase dream recall. It progresses to enhance dream color, dream interaction, and finally dream manipulation. Dream manipulation is when we can do work on ourselves, or intentionally visit with others, in dreamstate. Our spirit does not require sleep. Only our body and mind do. The application of Azurite is to hold it for about 5 minutes against the third eye just prior to falling asleep. The piece of Azurite must be free from the presence of Malachite (they often grow together) as the combination of the two is another tool altogether. Azurite connects the vigor to the subconscious, allowing it to promote self-healing. Your subconscious is the storeroom for all things not in the conscious mind. This brilliant blue stone gets its name from its azure blue color. It is comprised of copper carbonate hydroxide. When Azurite is combined with the Seven Cleansers it causes movement within the symptomology of an individual. This activity causes primary dysfunctional catalysts to move into a more prominent, or 'conscious', position for better observation and identification.

Azurite promotes self-healing explicitly in situations where diagnosis eludes us. When we do not know what is causing a given set of symptoms we use Azurite for its ability to seek out information. This is accomplished by the Azurite bringing primary symptoms to the forefront. As this happens, the nature of the actual imbalance is revealed. Clinicians are then able to draw accurate conclusions for diagnosis. It is then that we switch to the stone combination specific to the condition(s) identified.

As usual, using Azurite for other smooth muscle tissue will require the direction of other stones to identify where we are treating smooth muscle. Azurite is quite often found growing jointly with Malachite. Azurite is one tool, Malachite is another and Azurmalachite (See "The Endocrine System") still a third. You must be sure the stone used is of pure Azurite content.

Barite

A mold is always a fungus but a fungus is not always a mold. Indoor mold is not a source of infection. Barite, comprised mostly of barium sulfate, is for fungal infections. It fights off pathogenic fungi that infect those with compromised immune systems. Pathogenic fungi thrive off of living organisms. Barite treats tineas, ringworm, Candida, athlete's foot, nail fungus and yeast infections.

Fungal spores are often airborne and so are easily ingested into the lungs. Pneumonia, for example, can be the result of virus, bacteria OR fungi. If the pathogen causing the infection were fungi, Barite would be in your mix, as would the stones for lungs. Fungal infection may also be contracted by skin contact with soil containing the spores. Some infections are contagious and can be transmitted to other individuals. Infection can be spread by direct contact or by clothing, shoes and toiletries such as a hairbrush. Once again, it is a weakened immune system that allows the progression of fungal infection. Generally fungal infections begin in the lungs or on the skin and spread relatively slowly.

Some fungi are internal like Candida and yeast infections. A total combination would involve the Seven Cleansers plus Amethyst and Barite. This sentence says "fungal infection inside the body". If however the treatment is for tineas such as athlete's foot, jock itch and ringworm the combination changes. Since these are eruptions on the skin surface you would need to include Rhodocrosite indicating the involvement of epidermis. The addition of one stone, Rhodocrosite, changes the sentence to "fungal infection on the skin surface". Allergies and illness due to mold are not treated with Barite. They are addressed with Moldavite.

Bornite

Bornite, also known as "peacock ore", is for personality disorders. This copper iron sulfide is a metallic looking, rainbow-lustered rock, not to be confused with chalcopyrite. Personality disorders are innumerable. They are the presence of long term maladaptive patterns of behavior, cognition and inner experience, resulting in rigid patterns of thought and behavior. Bornite deals with the energetics of ego-syntonic. This is the medical term used to describe abnormalities in the ego and self-image. Currently, personality disorders are not treated with drugs so much as addressed with psychotherapy. The behavioral patterns in personality disorders are typically associated with severe disturbances in the behavioral tendencies of an individual, usually involving several areas of the personality, and are nearly always associated with considerable personal and social disruption.

Diagnostic criteria include evaluation of four thought processes: 1) The patient's cognition is the process of forming opinions about oneself and others. 2) Affectivity considers the range intensity and appropriateness of emotional arousal and response. 3) Control over impulses and gratification of needs, and 4) the manner in which one relates to others and handles interpersonal situations.

Additional indicators for the diagnosis 'Personality Disorder' are the following: The behavior is inflexible and pervasive across a broad range of personal and social situations. The pattern leads to clinically significant distress or impairment in social and occupational functionality. The behavior pattern is stable and of long duration. Its onset can be traced back to adolescence or at least early adulthood. It fails to meet the criteria of other mental disorders. And lastly, that the behavior is not due to a physiological response from injury or use of other substances such as medications, drugs or alcohol.

The ten recognized personality disorders are Antisocial, Borderline, Histrionic, Narcissistic, Avoidant, Dependent, and Obsessive-Compulsive, Schizotypal, Paranoid, and Schizoid (not to be confused with pineal disruption-caused Paranoid Schizophrenia),

These diseases of attitude contribute to the unbalanced social system known as patriarchal, where dominance and control rule.

Attitudes of self-entitlement are predominant. Matriarchal qualities, in contrast, include compassion and nurturing. Patriarchal leadership forces have done much to damage the human spirit.

Personality disorders are not gender specific. Men as well as women use the influence of Bornite. Since personality disorders are a learned response, the time treatment will take is significantly impacted by the source environmental trauma(s). In most, if not all cases, we will see the connection to childhood experiences. If the disorder can be linked to the age of pre-pubescence, Selenite will be included. Introspection and clarity are necessary for recovery. Unfortunately, the very nature of the illness, to not take responsibility for one's own life circumstances, makes the required candor difficult whether in self-dialogue or therapy.

Bornite is however, one of three stones that when combined are used as part of the formula for the male pituitary (See "The Endocrine System"). Women do not use the combination for male hormone balance as it relates to the total hormonal relationships within the male physiology and not simply on sex hormone levels such as testosterone (See "Selenite"). The other two male hormone stones are Citrine and Ulexite, which would be added to the Seven Cleansers and Amethyst.

Calcite – (Clear or White)

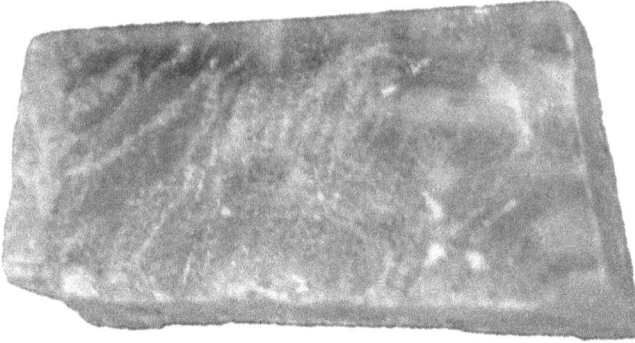

As calcium carbonate, is for mutated cell growth as cancer. Calcium carbonate triggers cell division. Hundreds of millions of living cells make up the body. Normally cells grow, divide, and die in an orderly fashion. Cancer is not as defined as you might think. It refers to diseases that display uncontrolled, invasive or destructive cell division.

This is the result of cells with damaged DNA continuing to replicate instead of die, as they should. Malignant cancers can metastasize and be carried to other areas of the body by the blood and lymph systems. It is an umbrella term assigned to any abnormally mutating cells. This can take the form of tumors or growths where they should not be. It also describes the mass degradation of cells.

Not all combinations used with Calcite will necessarily be classified as cancer treatment. All classifications of cancer, however, will be treated with Calcite. White Calcite is the general application stone for cancer anywhere in the body. It is important to know your stones, especially with these Calcites. White Calcite is used when there is no colored Calcite called for. There are many stones that could be combined with the white Calcite, Amethyst and the Seven Cleansers to denote location. Lepidolite would target the brain. Black and pink Tourmaline would indicate the lungs. Coral would mean that bone was compromised... add Aquamarine and the bone is the jaw, and so on. As you will see in the text that follows, the other Calcites target specific body parts for treatment as defined by their color.

Pink Calcite

Pink Calcite or, Calcite manganese is specifically for the synergistic repair of leukocytes. The white blood cells that normally fight off infection can cause blood cancer and other hematological malignancies. Some of them include myeloma, sickle cell anemia, Non-Hodgkin's lymphoma and leukemia.

Leukemia is a form of cancer that affects the body's blood-making system, including the lymphatic system and bone marrow. Leukemia involves the white blood cells, the leukocytes. Of the five primary types of leukemia, there are three that involve the lymph system. There are two other forms of leukemia that attack bone marrow. To give definition to these influences we provide the following combination examples.

With a base of the Seven Cleansers and Amethyst, we add pink Calcite, to indicate a blood-borne cancer, and blue Calcite to denote the lymph system. This stone formula spells out *a physical, white blood cell malignancy located in the lymph system.*

If we simply eliminate the blue Calcite and replace it with Iolite, the new message is the same as before except for location. Iolite redirects the other stones' energy to a blood cell malignancy in the bone marrow.

Yellow (or Orange) Calcite

Yellow (or Orange) Calcite is for use when directing healing towards the kidneys, liver, and bladders including the gallbladder and the *gland/organ*, the pancreas. The calcium carbonate seeks mutating cells, and the iron content (which turns it yellow or orange) defines the specific locale of the above-mentioned organs. (See "Endocrine System") This applies whether the diagnosis is specifically cancer or not. All these systems deal in some way with waste detection, processing and elimination.

The pancreas is an organ and a gland that produces various secretions including hormones. As such, it is a part of the endocrine system and is found in that section of this book. (See "Endocrine System")

In the kidneys, excess protein is broken down and eliminated in the urine. The liver is a filter for protecting our bodies from foreign substances such as is found in processed foods and medications. Bile production, animal starch storage and elimination of waste all take place in the liver. Many pharmaceuticals are linked to liver damage because that is where the body attempts to deal with them as waste. The urinary bladder collects urine and the gallbladder stores bile.

Green Calcite

Green Calcite is for the digestive system. The stomach, spleen, colon, bowels and intestines are parts of the digestive system.

The stomach receives food from the esophagus through the cardiac sphincter, which prevents food from backing up into the esophagus. Malfunctions cause acid reflux and related maladies. Green Calcite's influence begins at that sphincter.

In the stomach, food is broken down by the gastric juices. The stomach lining absorbs some nutrient matter. The remainder is processed by the stomach's five layers and passed on to the small intestine.

Tucked in at the upper edge of the stomach is the spleen. The spleen is a multitasker. It removes old blood cells as well as stores reserves. It recycles iron and produces antibodies, making it an essential part of the immune function fighting infections.

Digested matter enters the small intestine, which is responsible for absorption of nearly all nutrients into the blood. Improperly working, it is the catalyst for many serious and pervasive conditions including Irritable Bowel, Ulcerative Colitis and Crohn's Disease, occurring in the transport tubing of the large intestines.

Ultimately delivered to the colon, and its four sections, final waste is processed. The colon absorbs some fat-soluble vitamins. It is situated at the end of the digestive systems and extracts water and salt from fecal matter before it is eliminated.

This simple review of body functions is meant to assist with the ability to properly combine all stones needed for use. Only medically trained clinicians can give accurate diagnostics from which the formulas are derived.

Blue Calcite

Blue Calcite turned blue by niobium content in the mineral is specifically for the lymphatic system. The lymphatic system consists of organs, ducts, and nodes. There are no lymph glands. (The term "lymph gland" is a misnomer) Lymph nodes refer to the fabric of nodules running from neck to chin and armpits, including the mammary area, and a secondary system in the groin.

The lymphatic system is the drainage system for a clear fluid called lymph. This fluid is the result of blood plasma filtering through vessel walls then seeping into surrounding tissue. The lymphatic system gathers and transports this fluid back into the circulatory system.

Generally swollen when infection is present, the lymph nodes circulate blood components and manufacture antibodies. Lymphocytes produced fight off a multitude of infections, including many precancerous conditions.

As a system supporting our defenses against disease, Calcites will always be used for the treatment and prevention of cancers. This includes internal and external eruptions. Again note that the use of the various colored Calcites is not exclusive to cancerous conditions. They will be incorporated into combinations where the calcite serves to direct the stones to a specific area of the body.

For example, the treatment of breast cancer would be the following stone sentence: (7 Cleansers plus Amethyst)+ (blue Calcite)+ (Pearl)

(Physical)(Abnormal cell mitosis in the lymph system)(Specifically associated with the female reproductive system)

Chrysocolla

A copper based mineral, is what we use for the comprehension, retention, recall, and application of information. In other words, it is the processing of knowledge or 'input'. The basal ganglia plays a role in cognitive and memory tasks. It is also associated with motor abnormalities such as occur in Parkinson's, Huntington's and Wilson's diseases. This is also the stone of choice for any and all students. It will boost grades and performance in academic situations. Let us remember, though, that we are all always students in this life and can all benefit from the influence of Chrysocolla to help with converting our knowledge to the power of wisdom!

Chrysoprase

Chrysoprase is a chalcedony turned green by nickel content. Nickel in the body resonates with lipids, or fatty tissue substances. This is called adipose tissue and, when inflamed, causes pain. Pain is a signal from the nervous system outside the brain and spinal cord. It is our sensory nervous system's method of *touch* both interior and exterior. Pain can be vital in determining a diagnosis as it indicates the presence of illness or damage. Pain may be dull or sharp. It may be intermittent or constant. A jolt of pain might indicate injury or strain such as in broken limbs and pulled muscles. Or, it may be an ever increasing pain as with viral or bacterial infection. A clear diagnosis is always necessary before the elimination of pain is appropriate.

Once the cause of pain has been determined it is no longer needed. Pain following injury or incision is always expected. Pain is associated with a variety of illnesses ranging from influenza to cancer. Chrysoprase does not treat any given area of the body but serves only to block the discomfort of pain. It is for addressing the despair of pain. Pain weighs heavily on the mind, body and spirit

Chronic pain is debilitating. Those who suffer from it find it cripples their ability to function on a daily basis. Even the simplest of tasks become daunting to those who live with it. Chrysoprase puts a barrier between you and your pain so that function is restored while treating the underlying imbalance with other Rock-Medicine combinations.

Many circumstances arise where you will treat the condition and the pain separately. Arthritis would be treated with a specific combination of stones and a single piece of Chrysoprase used as needed for pain relief between applications. The same would apply to a broken arm for example, treated with the Seven Cleansers, Amethyst and Coral four times a day and the Chrysoprase used separately and individually, as needed for pain.

Pharmaceutical pain relievers slow down the healing process. Many over the counter and prescription pain medicines result in serious side effects including kidney and liver damage, not to mention addiction. Chrysoprase does not block subsequent warning pain from over exertion that may reinjure or impede healing, as many invasive herbal and pharmaceutical pain relievers do.

The most effective application is hand-held Chrysoprase for up to thirty minutes. This is an exception to the normal twenty minute limit. Holding just the Chrysoprase for pain without other stones simplifies the exchange. There is no movement of toxins since Jade is not present. Having only a single stone's resonance to interface with, the body and the stone can benefit from an extended duration of contact. After thirty minutes, put the Chrysoprase aside to clear. Another cleared piece may be immediately picked up and used for continued pain relief. It is still necessary to clear a piece of Chrysoprase for three hours before using that particular piece again.

Cinnabar

Of mercury sulfide, Cinnabar is for treating the heart. Mercury toxicity results in hypertension, a direct result of the heart's dysfunction. This stone directs healing to the physical location of the heart. It is useful in all cases of heart disease.

The average heart beats 100,000 times a day moving the equivalent of 2000 gallons of blood. Its four parts move in contractions that distribute our blood throughout the cardiovascular system. Of particular import is the blood the heart sends to and from the lungs for the exchange of oxygen.

Some heart problems may include angina, murmurs, holes, obstructions and valve damage. Heart disease is the number one killer in the United States, most often taking the form of CAD

(Coronary Artery Disease). This is the cause of most heart attacks.

Taking into account that we cannot separate the Spirit from Mind and Body, the implications of the "heart" being the most adversely affected part of our body is significant indeed. Cinnabar treats all three. Miraculous processes take place in the heart at all levels, affecting the human condition at all levels.

As Cinnabar is *extremely mercury toxic*, mercury poisoning could result if handled too much or ingested. Not surprisingly, the vulnerability of the human body from the ingestion of mercury from our environment and seafood are the effects of mercury poisoning on the heart.

When doing a hand-held application you will **_always_** use a piece of clear Quartz Crystal, 'set' (See Setting Stones) to the Cinnabar resonance, so as to insure that no mercury particulates are absorbed through the pores in the hand. (See Setting Stones p.44)

This is also a great example of why we make an essence with a 'set' crystal (See Setting Stones) and not the stones themselves. *You would **_never_** submerge a piece of Cinnabar in water and then drink it. It would not be clean, toxin-free medicine. It would contain mercury, and the potential to be lethal.*

A complete combination base for the heart would include the Seven

Cleansers, with Amethyst and Cinnabar charged crystal (see above). If you were to add Halite you would be saying heart muscle. If you added Garnet instead of Halite, you now are specifying a blood vessel in the heart. A combination of Cinnabar + Garnet + Coral indicates a calcium build up on the interior wall of a blood vessel in the heart. Learning the individual stones allows you to be amazingly precise with your stone combinations!

Citrine

Citrine is for itch and is one of the stones included in the treatment of the male endocrine system. Natural Citrine is not very abundant. Most of what is commercially available is actually artificially heated amethyst or smoky quartz. Citrine gems, versus citrine quartz, are more likely to be genuine. Be sure of authenticity of your citrine. The coloration of citrine is due to ferric impurities making it an Iron oxide compound. It performs like antihistamine for counteracting the itch response from various causes, which affect the peripheral nerves and spinothalamic tract.

When we say itch we are referring to the causes of a 'scratch' reflex. Many irritants provoke a scratch reflex or an itch. Various botanical secretions, insect bites and skin conditions can all cause an itch. Allergic reactions can also cause itching.

An itch, like pain, is another way that our bodies get our attention to an afflicted area. Once the cause of the irritant is identified the itch no longer serves a purpose and should be relieved. Similar to the way that Chrysoprase works on pain, Citrine treats the itch but not its cause. Where pain's discomfort is blocked by Chrysoprase, Citrine blocks the sense of irritation that itches promote. Citrine will not treat the cause of the irritant. Once determined, that must be addressed with an appropriate stone combination.

Citrine is a single stone application when it is applied for itch. It can be hand-held for up to thirty minutes when held alone.

Citrine is also one of three stones that combine with the Seven Cleansers and Amethyst for the treatment of the male hormone function being the endocrine system or, more specifically, the pituitary gland. The other two stones are **Bornite** and **Ulexite**.

Women never use the complete combination for male hormone

balance. And men never use the formula for women. The determinant factor is the presence of one X and one Y chromosome in the body cells. (Where the female is XX.) The hormone functions involved are much more than to regulate estrogen and testosterone. (See "Endocrine System")

Coral

Coral is to indicate calcium-based structures such as bone or any calcium based deposit. Calcium salts represent up to 70% of our bones and teeth weight. Coral is skeletal calcium carbonate formations left behind by the world's marine organisms that derive nearly all their nutrients from sunlight. Coral is waste product secreted by tiny polyps from the ocean and is now endangered due to environmental disruption and pollution.

Applied in all instances where calcifications occur, Coral could be considered a "what". Its presence says *calcium based*. Combining other stones for their specific local would tell the Coral "where".

Malachite states that the area to be addressed is the eyes. In the eyes a calcium based occurrence is a cataract.

Using Aquamarine would direct coral to seek calcium disruption in the mouth. In the mouth it would apply to the teeth and jaw bone. Emerald indicates the back. Coral would specify the back 'bone' or vertebrae in the spine.

You would not use more than one location stone at a time.

Without a specific location stone in the equation, Coral will address a broken or diseased bone anywhere else in the body. Coral is not used for the bone marrow. That stone would be Iolite.

Manual manipulation will always be a part of medicine. A broken bone often requires stabilizing before anything else. Once the bone is set, a combination of Seven Cleansers, Amethyst and Coral will be administered to expedite the knitting process.

Coral can be applied to all bone disruptions, breaks and calcium deposits on bone. Weak and brittle bones will be strengthened with Coral. The skeletal structure of any vertebrate is its base of support.

Emerald

Emerald is a combination of Vanadium, which helps regulate HDL in the spinal chord, and Chromium, and is for the spine. The spine is comprised of the Thoracic, Lumbar, Sacral, and Coccygeal vertebrae. Combining other stones denotes muscles, bones, nerves, joints, and even fluids. Chronic back problems are predominant in our society worldwide; however, the causes differ.

Problems from injury may involve discs, muscles and/or vertebrae. ALS, or "Lou Gehrig's Disease," is a progressive neurodegenerative disease that affects nerve cells in the brain and the spinal cord. Motor neurons reach from the brain to the spinal cord and from the spinal cord to the muscles throughout the body.

The progressive degeneration of the motor neurons in ALS eventually leads to their death. When the motor neurons die, the ability of the brain to initiate and control muscle movement is lost. The treatment would include Emerald for the spine and Rhodonite to indicate the specific area of motor neurons.

Cervical vertebrae C1 through C7 are considered the neck, and so Herkimer is used instead of Emerald to indicate that distinction. The back is like a zipper running from the base of the skull down to the tailbone. When one portion is out of balance it can throw the entire back out of balance.

Our back is the primary skeletal support for our physical structure. Without it, we would be like a water balloon on the floor. Back issues cover a very wide spectrum, from bad posture to stress pockets. Emerald serves as the "where" in your equation. In the following scenarios, the back is the 'where'.

If the 'what' is muscle, the stone equation it would call for Halite. The combination of Seven Cleansers with Amethyst, Halite and Emerald makes a resonant statement. It says; "We are to bring balance to compromised **muscle** matter located in the back".

Changing just the Halite out, to a piece of Coral, completely changes the message. The new resonant statement says; "We are to bring balance to compromised **bone** matter located in the back". If the damage is neurological, as with pinched, bruised or severed nerves, then Rhodonite is used as the 'what' in the combination. It specifies **nerve** damage or disruption. Consider also the stones such as Rhodocrosite that says skin, and Sulfur for inflammation of connective joint tissue.

The success of Rock-Medicine is wholly dependent on precision combinations that do not send mixed messages. You cannot treat muscle and bone in the same combination, but you can combine Halite with Coral with Sulfur and you are describing "connective joint tissue between muscle and bone which is a tendon. Without the Halite, it is connective joint tissue bone to bone, or a ligament. It must be one or the other. A complete combination of Seven Cleansers with Amethyst, Emerald and Coral would address back bone only. It could be immediately followed with another complete application of Seven Cleansers, Amethyst, Emerald and Halite for back muscle. A general back application could be Seven Cleansers, Amethyst and Emerald. Know your stones!

Fluorite

Fluorite occurs in many colors. It is calcium fluoride, and in the human body the vibration of this mineral acts as flux. Edemas, or abnormal accumulations of fluid, can occur in patients with disease of the heart, the liver, or of the kidneys, and in the victims of malnutrition. In medical terms it is the discharge of large quantities of fluid material from the body. All colors work the same for Rock-Medicine. Fluorite is for the presence of unwanted fluid build-up. There are five categories of bodily fluid, and include everything from semen to earwax. Due to the naturally dense hydration level of organic life, the human body is a soup of ever-changing fluids of blood, water, waste, and chemical processes.

Fluid build-up can be the result of many circumstances, from injury to illness. Fluorite is the 'what' in your sentence. The stone(s) added for direction will determine 'where' the fluid build-up is.

Many stones are offered to give assistance to specifically 'where'. The Seven Cleansers with Amethyst, Lepidolite and Fluorite would indicate fluid on the brain.

Excessive accumulations of fluid on and around joints, like the knee and elbow, would call for the Seven Cleansers with Amethyst and Sulfur.

Black Tourmaline (with pink) directs the resonance to the lungs for pleurisy or pulmonary edema. Peripheral edema is the Seven Cleansers, Amethyst and Fluorite. There is no additional combination stone known to specify the lower legs, ankles and feet.

Fluid build-up could be causal or symptomatic and can occur virtually anywhere in the body. When there are multiple conflicting locations involved, or where there is no specification for location by additional stones, the Seven Cleansers with Amethyst and Fluorite would be a generic, whole-body application. Ruptured discs in the back are the result of the gelatinous nucleus of the disc pushing through its hard casing. It would be treated with Fluorite, along with the Seven Cleansers, Amethyst and Emerald indicating a fluid problem in the back.

Galena

Galena is for respiratory rate and the system that supports it. This system is comprised of the trachea, bronchi and diaphragm. Infections can include viruses, bacteria, fungi and parasites. Galena with Bloodstone would indicate a virus attacking the respiratory system. Galena with Iolite would indicate a bacterial invasion of the respiratory system. Galena with Larimar would treat parasitic protozoa, if it were to involve the respiratory system...and so on. Galena is silver in color and box-like in crystalline structure. It is a lead sulfite. When someone suffers from lead poisoning it is the oxygen-carrying protein in the hemoglobin that is targeted and harmed. This is why lead residue from the stones can pose a threat and we must be careful when handling Galena. Small stable pieces are recommended for hand-held application

Our body rhythms are called our 'vital signs' and the respiratory rate is one of the most closely monitored during illness. Galena focuses on the apparatus and rhythm of breathing. The number of breaths, or inspirations and expirations in a time unit measures rate.

Adults breathe on average anywhere from 13 to 45 breaths per minute depending on exertion. Every breath we take fills us with stimuli and chemicals...our lungs process the gasses on their inner surface... and we exhale. This process is repeated for each of the 1200 or more breaths we take per hour.

The primary exchange happens in the alveoli. With each inhalation, oxygen is extracted and diffused into the hemoglobin. Simultaneously carbon dioxide is removed and exhaled.

When there is impairment due to structural abnormalities, Galena will combine with other stones to specify epithelial, or vessels, or muscle. In this manner we can treat the apparatus, or form, verses function. Including Galena with just the Seven Cleansers and Amethyst targets the respiratory rate.

Respiratory rate can fluctuate due to respiratory problems including allergies, asthma, bronchitis, emphysema, and hyperventilation caused by panic, or extreme physical exertion. Respiratory rate is affected by both emotional and physiological responses.

Many respiratory episodes are triggered by stress, whether over the short or long term. Galena helps regulate that rhythm and process. Exercise, relaxation and meditation are prevention practices that we can do to balance our spiritual, mental and physical bodies' rhythm.

Garnet – Red

Garnet is Iron aluminum silicate, and is for the circulatory system. Also known as the vascular system, this is the delivery system for a host of blood-borne chemicals and nutrients throughout the body. The circulatory system is the body's network of blood vessels including arteries, veins and capillaries. Garnet alone says any and all of these are targeted.

Arteries and veins differ from microscopic blood vessels called capillaries. The primary difference is a layered muscle membrane encasing them. The combination of Garnet + Halite would indicate blood vessel + muscle. All cardiovascular disease treatments will include Garnet and Cinnabar.

Garnet will be used for blood pressure problems. The function of the circulatory system is what we measure for blood pressure. This is another 'vital sign'. There are two energy actions that combine for blood pressure. One is the force used by the heart to pump blood into the circulatory system. The other is the force with which the arteries resist the blood flow. Because the heart generates the force, we add Cinnabar. Cinnabar with Garnet concentrates the stones' focus on the blood vessels in and attached to the heart. Add Rhodocrosite for epithelial, and you are treating the thin epithelial lining of a blood vessel itself.

Blood pressure readings are measured in millimeters of mercury and typically given as 2 numbers. For example, 110 over 70 (written as 110/70).

• The top number is the systolic blood pressure reading. It represents the maximum pressure exerted when the heart contracts.
• The bottom number is the diastolic blood pressure reading. It represents the minimum pressure in the arteries when the heart relaxes.

Seven Cleansers with Amethyst, Garnet and Cinnabar will be the application for both high and low blood pressure. High blood pressure is dangerous to the heart because it makes it work harder.

In as many as 95% of the cases there is no clear determining factor as to cause. It is known as hypertension.

Low blood pressure is less of a threat. It is more significant to see how quickly the rate drops as opposed to how low it goes. Again, as with hypertension, there are numerous conditions, and even situations, that can cause fluctuations in blood pressure. Where these other areas of involvement are defined, additional stones may apply. *Please read precautions regarding Galena and Cinnabar prior to use.*

Rock-Medicine is not contraindicated for use with any other treatments or medications. It is essential, however, to have the levels checked every couple weeks once on Rock-Medicine. As the body heals, the need for those medications will be diminished and eliminated.

Halite

Halite is a rock salt and is for muscle. Salt is essential for muscle contraction and mass. This is a *'what', or 'muscle'*. Many stones are offered to specify the *'where'*. There are three categories of muscle in the body. They are skeletal, cardiac and smooth. Halite addresses the skeletal muscle group only. Azurite addresses the smooth muscle group. As the heart is itself a muscle the use of Cinnabar is its sole identifier.

The skeletal muscles are for locomotion, such as the quadriceps in the thighs. This category also includes those muscles for movements, such as are found in the eye.

There are three types of skeletal muscle damage. They are tears, strains and loss of elasticity. Muscles can be compromised by way of injury or disease. Some pharmaceuticals, such as statins for cholesterol, include muscle impacting side effects.

All forms of physical therapy, chiropractic and sports medicine are benefitted by the inclusion of a Rock-Medicine combination containing Halite. So, too, is any type of 'range of motion' exercise used to prevent or address atrophy. This stone is an excellent preventative measure for geriatrics, and those diagnosed with degenerative muscle disease.

Halite is another fine influence to use on pregnant women and infants to give balance to the foundation of muscle development. Halite is excellent for use by athletes to minimize cramps and rigidity. It provides protection against greater degrees of injury. This applies to people who work strenuous physical jobs. Dancing, gymnastics, aerobics, running, jogging, calisthenics...all activities that Halite protects the muscles during the use of. It assists in healing the effects of strain or injury during participation in all types of physical activity.

Building muscle is a key element in reducing body fat. Weight loss programs, especially through vertical banded gastroplasty, should include an application for muscle restoration. And use a separate combination for restoring skin elasticity including Rhodonite. Halite is used, for example, to treat Muscular Dystrophy.

Included in a combination of Seven Cleansers with Amethyst and Malachite, Halite would specify that the muscle to be treated is located in the eye. The six eye muscles are called 'extraocular muscles'. They control the directional movement of the eye. Where eye problems exist as a result of the eye muscle's inability to rotate or move the eye, Halite will be part of the combination.

As a salt, Halite is water-soluble. Always use a 'set' crystal (See Setting Stones) for rendering essences. A set crystal is used in a hand-held application to avoid it melting into the pores of the palm, as it is a salt.

Herkimer Diamond

Herkimer Diamond is for the neck and throat. Herkimer is a type of double terminated quartz that is formed in a host material of mostly magnesium. The anatomy of the throat contains a variety of multipurpose systems and magnesium is closely associated with many. This includes the larynx, vocal cords and esophagus. These parts all have their own function but combine to make the throat.

Larynx Functions

- Control of the airflow during breathing
- Sound creation
- Prevention of air from escaping the lungs, for example, during

weightlifting

• Protection of lungs against foreign objects

Vocal Cord Function

• The vocal cords are a set of membranes that stretch over the larynx. They have only one function and that is to allow speech.

Esophagus Function

• The esophagus has only one important function in the body—to carry food, liquids, and saliva from the mouth to the stomach.

The neck is comprised of muscles, bones, joints and nerves. In stones, this list reads as Halite, Coral, Sulfur, and Rhodonite. With a base of the Seven Cleansers and Amethyst, combination stones are added as to complete the description of the disease.

Some equations for addition appear as follows. The addition of Herkimer and Amethyst specifies neck muscle. Esophageal cancer would include Herkimer and white Calcite. If you use Herkimer with Coral and Calcite you would be describing any bone cell mutation occurring in C1 to C7, whether it is malignant or benign. This can include infection.

Iolite

Iolite is magnesium, iron and aluminum. Iolite is for bone marrow. Our blood is formed in our bone marrow. The other primary function of our bone marrow is the formation of Phagocytes. These are responsible for fighting off bad bacteria and are the soldiers of the immune system. They are responsible for swallowing, killing and digesting invading bacteria. Some bacterial infections include; pneumonia, ear infections, diarrhea, urinary tract infections, and skin disorders.

Bone marrow is spongy tissue inside most bones. There are two forms; *yellow marrow,* consisting mostly of fat cells such as is found in thigh and hip bones and *red marrow,* formed in the flat bones, which contains immature cells known as stem cells. All our marrow is red at birth. As we get older more and more yellow is produced. Both types of bone marrow contain numerous blood vessels and capillaries.

Iolite alone impacts the connective yellow marrow in the hollow center of the long bones. Yellow marrow is involved in fat storage. Yellow marrow can also convert back to red marrow in a crisis. It produces lymphocytes that modulate the functional activities of many other types of cells.

Red marrow is where our body creates our blood and antibacterial phagocytes. The stem cells develop into the red blood cells that carry oxygen through your body, the white blood cells that fight bacterial infections, and the platelets that help with blood clotting. Adding pink Calcite specifies the red marrow produced in the flat bones. This explains why pink Calcite would combine with Iolite in the treatment of leukemia, for example.

Ivory

Ivory is for the male reproductive system consisting of the prostate, penis, scrotum and testicles (testes). Ivory is predominantly made up of collagen. We can use any tusk type ivory. As with shark's tooth, we do not kill an animal for its ivory. Both fossilized and new ivory are acceptable for use.

Females do not use Ivory as males do not use Pearl. Unlike women's, parts of the male reproductive system are located outside the body cavity. These are the penis and portions of the scrotum including the testes. A collagen membrane surrounds the testicles.

The prostate is more vulnerable than some other areas to the development of cancer and other disease. It produces and regulates various fluids, including semen. Since it is a gland, when treating the prostate, we will always include Bornite, Citrine and Ulexite with the Ivory. (See "The Endocrine System") If it is cancer to be treated, we add Calcite.

So, a prostate cancer sentence incorporates 13 stones and looks like this;

(7 Cleansers + Amethyst + Calcite) + (Bornite + Citrine + Ulexite) + (Ivory)

It interprets as;

(There are malignant cells) (In a male gland) (Located in his reproductive system)

Infertility in men is caused by a wide variety of conditions. It can be emotional blockage or a physiological disruption. It could be anything from poor sleep to poor diet.

In the case of previous damage due to infection or injury to the reproductive systems, we use the Seven Cleansers with Amethyst plus Ivory. STDs and PIDs are responsible for about 40% of infertility problems in both genders. Dealing with scar tissue will take a longer period of time to respond than with more active tissue.

Larimar

Larimar is a form of pectolite. Common pectolites are formed of sodium calcium inosilicate hydroxide. The blue pectolite, Larimar, gets its color from the calcium being replaced with cobalt. Its unique function is the eradication of parasitic protozoa. The list is long but some of the more prevalent pathogenic protozoan illnesses are:

Malaria
Amoebiasis
Giardiasis
African Sleeping Sickness
Leishmaniasis
Toxoplasmosis
Babesiosis
Trichomonia

Use by combining it with the Seven Cleansers and Amethyst.

Lepidolite

Lepidolite indicates the locale of the brain. Lepidolite is rich in lithium. Lithium promotes the health and development of brain neurons. Protecting neurons from destruction is one thing; it's quite another to stimulate the growth of new neurons—a process called *neurogenesis.* The highly complex human brain is the center of the nervous system. An adult human cerebral cortex contains as many as 35 billion neurons, *each* linking up to several thousand synaptic connections. The brain is an amazing bio-chemical computer. It is the receiver of all education, source of all behavior, and prompt of all response. Everything affects the brain and the brain affects everything.

The cortex is divided into four lobes. The frontal lobe, where you do your heavy thinking, pondering and planning your actions; the temporal cortex, where you process sounds and form memories; the occipital cortex, where you process all the things that you see; and the parietal cortex, where you integrate, or make sense, of all of the different bits of information that are bombarding your brain. The brain structure is easy to see. Its function is quite another science.

The brain is so complex and intricate that we can only see its form and the result of its function. There are those who believe, for example, that patients who suffer from dementia are in fact cognizant in their mind and unable to convey messages for communication. They are trapped inside their own head. Such is the description of many mental illnesses. Some are indeed chemical imbalances and are able to be treated with combinations, including Lepidolite. Do not confuse personality disorders with mental illness. (See Bornite)

Lepidolite is the 'where' in our sentence. If used alone it addresses general chemical imbalances in the brain. The stone itself is the mineral form of lithium. A general brain chemical treatment of Seven Cleansers, Amethyst and Lepidolite has a wide spectrum of influence. It is used for bipolar and clinical depression. It repairs damage from prolonged drug and/or alcohol abuse and corrects the recessive gene contributing to the syndrome.

When the imbalance is defined by the brain's neurological system, Rhodonite (neurological) is added to the Lepidolite (brain) indicating the central nervous system.

If the condition is fluid on the brain, Fluorite (fluid) is used with the Seven Cleansers, Amethyst and Lepidolite (brain).

If the diagnosis is cancer on the brain, the Seven Cleansers with Amethyst, Lepidolite (brain) and white Calcite (abnormal cell division) combine for treatment.

We are barely touching on all the possibilities for brain disease, especially since this is where the body meets the mind. The only way to fully use Rock-Medicine is to know the individual stones and what they represent.

Malachite

Malachite, copper carbonate, is for the eyes. The eyes react to light. They capture information at a cognitive level. Sight is one of our primary senses.

Optometry is the branch of healthcare dealing in eye care and vision. It is an entire course of study unto itself. Ocular medicine is intricate and complicated. There are thirty different components to the eye; some tissue, some vascular, some fluid, some muscle, even some space.

Following is a list of eye parts and their correlation. This is the first time there has been an attempt to clarify each and every eye parts' relationship to individual stones as opposed to general locale. A clinically trained physician will need to evaluate the nature of eye conditions that present, and the subtleties of the mineral influences as they apply.

Posterior chamber = space
Ora Serrata = vascular
Ciliary Muscle = muscle
Ciliary Zonules = vascular
Canal of Schlemm = fluid canal
Pupil = space
Anterior Chamber = fluid
Cornea = epithelium
Iris = muscle
Lens Cortex = tissue
Lens Nucleus = epithelium
Ciliary Process = vascular
Conjunctiva = epithelial
Inferior Oblique muscle = muscle
Inferior Rectus muscle = muscle
Medial Rectus muscle = muscle
Retinal arteries and veins = vascular
Optic Disc = nerve
Dura Mater = nerve
Central Retinal artery = vascular
Central Retinal vein = vascular
Optic nerve = nerve
Vorticose vein = vascular
Bulbar sheath = connective tissue
Macula = cell division
Fovea = cell division
Sclera = tissue
Choroid = connective tissue
Superior Rectus muscle = muscle
Retina = nerve

One of the most common conditions of diminished vision is macular degeneration, in particular age-related macular degeneration (AMD). This is a condition of the mutation of blood vessel cells between the retina and the choroid. Increased vessel matter pushes against the retina and can cause it to detach. The retina is responsible for clarity of vision. The retina is the nerve surface that the eye uses to display what it sees to the brain. Damaged retinas inhibit the clear reception of images viewed. This and similar conditions are gene mutations, making them hereditary. AMD is degenerative.

The application to eliminate the tissue that is increasing vessel pressure would include the Seven Cleansers, Amethyst, Malachite and white Calcite. Untreated, retinal integrity will continue to deteriorate, ending ultimately in the loss of all vision. Note that with Calcite we are using the stone sentence of "we are treating for balancing abnormal cell mutation in the eye".

The Retina is a nerve. If it were the target, the application would call for Rhodonite instead of white Calcite. Once the abnormal vascular growth is addressed with the inclusion of Calcite, any remaining damage to the retina itself would be addressed by Rhodonite with Seven Cleansers, Amethyst and Malachite.

Halite indicates muscle. Rhodonite denotes nerve. There are six extraocular muscles. Comprehending the relationships between Rock-Medicine and the intricate systems of eye muscles, nerves, fluid, vessels, tissue and voids will require a precise understanding of the stones associated with them. Glaucoma is the improper distribution of fluid around the eye. It causes increasing pressure. The Seven Cleansers with Amethyst, Malachite and Fluorite spells out the strategy for treatment.

A very general eye application for the eye would be the Seven Cleansers with Amethyst and Malachite. Review all stones for their application, and be sure not to send mixed messages within your final combination. Due to the complexity of ocular research, and the early stage of research that we are at with Rock-Medicine, we are only able to scratch the surface of information for the stones as used for the eyes. Most of our experience to date has included cataracts, infections and focus. Rock-Medicine as applied to eye care and vision will need to be an independent course of study within the currently existing Optometry Profession.

Moldavite

Not themselves extraterrestrial, but rather formed by the impact of meteorites, Moldavite is predominantly silica glass, magnesium and iron. Moldavite is for mold. Where Barite addresses pathogenic fungi (See "Barite"), Moldavite responds to **toxigenic mold**. Toxigenic mold is the type of fungi that grows in moisture and is toxic to all who come in contact with it. The danger is from mycotoxins. Clinical Environmental Mycotoxicology is also in its infancy. Mold as relating to mainstream health issues has only been researched within the past 10 years or so.

Mold needs only moisture and an organic food source such as wood or sheetrock in order to grow. Rock-Medicine cannot be used to clean mold from surfaces or kill its spores. It treats illness contracted by those who come into contact with it. Even healthy individuals can be made ill by mold and mildew. Those with allergies and other respiratory conditions are at higher risk.

Moldavite is used with the Seven Cleansers and Amethyst to counter toxic effects from contact or ingestion of molds and mildew.

Mother of Pearl

Mother of Pearl is a form of calcium carbonate. Mother of Pearl is to treat a fetus in the uterus. The name 'Mother of Pearl' came from the Latin word 'mater perlarun'....'of their mother's womb'. Only the presence of a fetus triggers the influence of Mother of Pearl. After a day or so of development, a fertilized egg travels the fallopian tube down to rest in the uterus. This embryonic home is our first food and shelter.

Early miscarriages are usually the result of an abnormal fetus. First trimester miscarriage can be due to chromosome defects, immune disorders or anatomically displaced fetus. These types of pregnancies are sometimes not even detected. Often the fetus is gone before any problem is identified.

In the second trimester primary threats to the pregnancy are from uterine malfunction. This may be either incompetent cervix or membrane failure, triggering, among other situations, early labor. About 25% of late miscarriages are due to incompetent cervix. In circumstances where it is the integrity of the womb that is failing, we use Pearl with the Seven Cleansers and Amethyst. Absent a pregnancy, disease of the uterus is treated with Pearl instead of Mother of Pearl.

There are many other conditions that are relative to the fetus itself and not the womb, which are addressed by other stone combinations. The precision with which we can control the stone message by combination is impressive. If we combine the Seven Cleansers with Amethyst and Coral we are specifying that healthy bone be supported and/or damaged bone repaired in a person. If we add Mother of Pearl we are telling the rest of the stones that the bone to be influenced is in the womb. If Lepidolite is used with Mother of Pearl the target area is "the brain located in the womb". Virtually any fetal abnormality can be treated with Rock-Medicine *in utero*. If there is no fetus the stones will not change direction and treat the mother. When any combination resonance has Mother of Pearl included, it is going to look in the womb for its assignment.

Many genetic fetal syndromes can be now detected early. We already do surgery on fetuses for organ malfunction. Rock-Medicine enables us to treat fetuses as precisely for healing and prevention as we treat ourselves.

The incorporation of Mother of Pearl with Selenite, among other stones, can be used for effective posttraumatic treatment of physical and emotional energy blockage sustained in utero. One example would be Autism.

Pearl

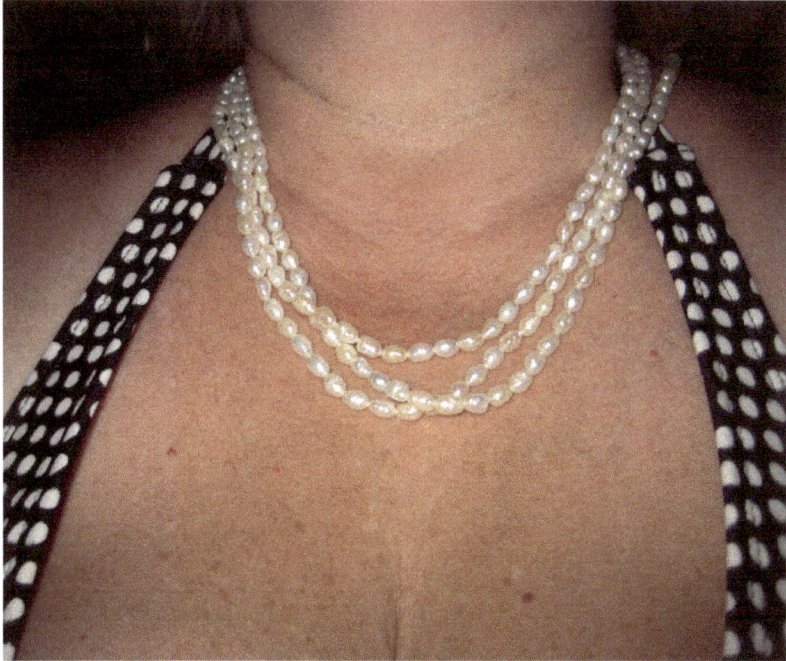

Natural pearls are formed randomly and really are simple accidents of nature. When a certain type of irritant, such as a parasite, becomes lodged in the tissue of a mollusk, the animal responds by secreting a calcium carbonate substance called nacre to coat the intruder and protect the mollusk. The nacre is concentric layers of minute calcium carbonate crystals, and forms a pearl.

Pearl is for the ovaries, fallopian tubes, uterus and vagina. It is included in the triad for the female endocrine system with Opal and Lapis. Only women use Pearl and Mother of Pearl. Eggs released monthly by the ovaries have the potential to be fertilized by sperm at precisely the right time so as to produce a new life. Conception is random. Once fertilized, the dividing cells travel through the fallopian tubes to settle in the uterus for the next nine months.

Other than producing eggs, the ovaries manufacture the female hormones estrogen and progesterone. Estrogen production in girls is one of the signals of the onset of puberty.

Ovarian cysts are fluid filled blister-like pustules that can easily occur and disappear. They are treated with Seven Cleansers, Amethyst, Pearl and Fluorite. Polycystic Ovary Syndrome (POD) is a much more chronic and common presentation of cysts that are triggered by a hormonal imbalance. The treatment would be the Seven Cleansers, Amethyst and Pearl. Long term effects from this progressive disease can be quite serious. The Seven Cleansers, Amethyst, white Calcite and Pearl would treat ovarian cancer.

In the case of previous damage due to infection or injury to the reproductive systems, we use the Seven Cleansers with Amethyst plus Pearl. STDs and PIDs are responsible for about 40% of infertility problems in both genders. Dealing with scar tissue will take a longer period of time to respond than will more active tissue.

The fallopian tubes are most often damaged by infection. Fallopian tubes grab an egg released from the ovary and inside the tube it is fertilized. The tube then serves to transport the fertilized egg to the uterus. If the tube is blocked or malfunctioning, conception is impossible. It is the fallopian tube most often to blame for the inability to conceive.

The uterus serves to facilitate a growing fetus. It has only the one function. Upon birth the child travels out through the birth canal or vagina.

Whether pregnant or not, the female reproductive system is heavily assaulted by pathogens. Reproductive parts can be infected, injured, displaced or contain tumors, both malignant and benign. All parts of the female reproductive system, except for the mammary glands, are represented by Pearl. Uterine cancer for example, would be treated with a combination of Seven Cleansers, Amethyst, Pearl and white Calcite. But, so would ovarian, fallopian, or vaginal cancer all be treated with the same combination. Bacterial infections would add Iolite. Viral infections would add Bloodstone, and fungal conditions would require Barite.

Rhodocrosite

Rhodocrosite is a high content carbon enriched by Manganese, which is the required co-factor for an enzyme called prolidase. Prolidase is necessary to make collagen as a structural component of skin. As an anhydrous carbonate calcite, the chemical vibration of Rhodocrosite serves to transport ions that regulate epithelial cells. Epithelial tissue is comprised of tightly packed cells that cover and line all internal and exterior surfaces of the body. Inside the body these epithelial layers are called endothelium. The top five layers of our outer skin are known as the epidermis. Rhodocrosite is used to signal that epithelial cells are involved in the ill or damaged area.

As we continue to say, some stones say 'what', some say 'where'.

Using the Seven Cleansers and Amethyst as the standard base, we are going to discuss some combinations and how they help pinpoint target areas and conditions. With Rhodocrosite added to your grouping, the equation now specifies that the condition is on the skin. If we use white Calcite in our sentence, we are indicating the condition consists of abnormally mutating cells such as cancer. This would apply to melanoma, for example.

If we start again, adding Rhodocrosite to the base, we are indicating the involvement of epithelial cells. Adding Cinnabar says those epithelial cells are in the heart. The endothelium is the epithelial lining of the arteries of the heart. Damage to the endothelium is called Atherosclerosis and is most often caused by smoking, high blood pressure and high cholesterol. Atherosclerosis is the primary cause of heart attack, stroke and peripheral vascular disease.

Another case example is a disease called Sacrococcygeal Teratoma (SCT). This is where a fetus has a tumor located on the coccyx, or tailbone. Some are tissue filled and may be malignant or benign. Using the standard base, we add Rhodocrosite to indicate there is a tissue growth. Emerald says it is in some part of the spinal column. Mother of Pearl tells the other stones it is in the womb. If the tumor is benign your combination is complete. If, however, it is malignant you will add white Calcite, indicating abnormal cell growth becoming progressively worse. Some SCTs are fluid filled. If we exchange the Rhodocrosite for Fluorite we have described the same scenario except that now the stones are all looking for a cystic, or fluid filled tumor, instead of a tissue filled one.

It can't be stated enough times that knowing what each and every stone does is the only way to fully access the Rock-Medicine system.

Rhodonite

Rhodonite is for the nervous system. Rhodonite is a manganese inosilicate. The body's nervous system is the most at risk from magnesium imbalance. Made up of neurons, our neurological cells do one thing that other cells do not, they communicate. The neurons' specialty is transmitting, not matter, but *information*. The element magnesium's ability to carry multiple charges simultaneously is vital to the function of the nervous system.

The nervous system is the body electric. Our neurological systems process *all information* and communicate it to the body, the mind, and the spirit. A nerve cell membrane has a positive charge on the exterior and a negative charge on the interior. Skin, blood, bones, and membranes are all things we can identify and measure. It is the neurological system that deals in chemicals leaping over empty space that is the perfect balance of an electrical charge's at the ready. Positive and negative ions can pass over the membrane and the electrical charges will be moved along the nerve cells. As with many other medical specialties, the neurology of our physical bodies is an extensive field of study all on its own.

The neurological system is divided into two different classifications. The **Central Nervous System** (CNS) coordinates all information exchange between the parts of multicellular animals, all of which have a central nervous system. Vertebrates are the most advanced life forms on Earth. Humans are the most advanced of the vertebrates. The human brain uses 100 billion neurons to communicate with the spinal cord. The central nervous system is what makes the brain a computer. The CNS can be disturbed by emotional trauma as well as physical disease.

Insomnia, hyperactivity, anxiety, panic attacks and most phobias are symptoms of magnesium deficiency affecting the central nervous system. Magnesium disruption is linked to some forms of dementia, which do not include hallucinations or delusion. A general application for the central nervous system would be comprised of the Seven Cleansers, Lepidolite and Rhodonite. Amethyst is notably absent here, as the units of function in the CNS are comprised of electrical charges and not dividing cells of matter. In the applications for infection, the Amethyst is included because there is infectious matter present and not for nerve cell division. (See "Amethyst")

The **Peripheral Nervous System** (PNS) is comprised of *sensory* and *motor* neurons connecting our spine to our limbs. The sensory neurons carry impulses towards the brain. The motor neurons carry impulses away from the brain. A stimulated nerve ending shoots data that makes an electrical inquiry of the brain. The brain receives the data, formulates and delivers a response. The process is virtually instantaneous.

The motor neurons control the brain's ability to move our extremities. When we touch a hot surface the sensory neurons inform the brain. It is the brain that tells our hand to recoil. If the communication path is damaged, the hand has no signal to react, whether the disruption is due to injury, infection, or the introduction of particular drugs. The peripheral nervous system responds to magnesium imbalance with various symptoms including numbness or tingling.

The PNS includes two other subsystems. The **Cranial** nervous system connects your brain to the sensory organs of the ears (Wood Opal), nose (Shark's Tooth), mouth (Aquamarine) and eyes

(Malachite). Eye movement is the result of the cranial nerve controlling ocular muscles. (The autonomic nerve controls Pupil dilation and constriction.)

The **Autonomic** Nervous System (ANS) connects the brain and spinal cord with your organs. It includes the hypothalamus connecting the nervous system to the endocrine system by way of the pituitary gland (See "Endocrine System"). It also is for metabolism, which is defined by the inclusion of Apatite (See "Apatite"). The ANS regulates heart rate (Cinnabar), digestion (green Calcite) and respiration rate (Galena). Whereas most of its actions are involuntary, some, such as breathing, work in tandem with the conscious mind. This extensive system dictates salivation (Aquamarine with Fluorite), perspiration (Rhodocrosite and Fluorite), and sexual arousal (Ruby). With control over temperature, sleep, water metabolism, pituitary gland, blood pressure, hunger, sympathetic and parasympathetic balance, the autonomic nervous system is of immense importance.

The Peripheral Nervous System is much more exposed than the well protected Central Nervous System, making it much more vulnerable to injury and infection. As stated, the central and peripheral nervous systems facilitate the entire body's communication. Combine the stones in a precise way.

Shark's Tooth

Shark teeth are made up of natural fluoride. One of the areas fluoride toxicity compromises is the nasal cartilage and olfactory systems. As such, Shark Teeth designate the nose and/or sinus area when used in a sentence. Just as with other stones that designate a 'where', the additional stones you add will serve to indicate 'what'. If mucus were present you would add Fluorite to the mix to specify excess fluid in the sinuses. If one had melanomas or other tumorous eruptions inside or on the nose, the combination would include white Calcite and Rhodochrosite. Sinus infection, whether viral, bacterial, or fungal, would respectively include Bloodstone, Iolite, or Barite. Since there is no actual bone in the nose, there would not be a correlation for adding Coral.

Numerous issues can affect our sense of smell. You will need to get a precise diagnosis in order to ascertain what contributing factors may be present. This can be a brain chemical origin or even a blockage in a subtle portion of the endocrine system. Once established, select the stone(s) that describe the function involved. In the whole of the animal kingdom it is said that the shark has the keenest sense of smell.

Sulfur

Sulfur is used for connective joint tissue. Sulfur is the third most abundant chemical in our bodies. We have two types of connective tissue in our joints, which are tendons and ligaments. Tendons connect muscle to bone. When specifying the target for healing is a tendon, we use Halite with Sulfur and Coral as part of the overall combination. When the target is a ligament we use Sulfur with Coral and leave out Halite, which denotes muscle. The cross-linkages through disulfide bonds strengthen the tissues that make up the joint.

Although bursitis and other arthritic conditions are the result of a compromised immune system, sulfur is an integral addition to all stages of combating the symptomology and damage that has been done. I would always recommend two applications. One of the Seven Cleansers with Amethyst to fix the immune system and a second with the Seven Cleansers, Amethyst and Sulfur to repair the joint tissue.

Tiger's Eye

Tiger's Eye is comprised mostly of magnesium silicate. It is a golden to red-brown color. Do not confuse it with red stones that are heat treated, or dark stones having had their colors artificially lightened using nitric acid. Nor an incompletely silicified blue variant called Hawk's eye. A natural Tiger's Eye is for fatigue. With the Seven Cleansers and Amethyst it can be used to address Chronic Fatigue Syndrome. For general fatigue or monotony, it may be held alone for 30 minutes, as is the case with Chrysoprase for pain, and Citrine for itch. Don't confuse general fatigue, with Chronic Fatigue Syndrome, which has been linked to a virus attacking the white matter of the brain.

Tourmaline

Black Tourmaline

Black Tourmaline, or schorl, is the most common of the tourmalines. It is a basic sodium iron aluminum boro-silicate and indicates the target area as the lungs. As a general cleansing for the lungs you would use it with the Seven Cleansers, Pink or Watermelon Tourmaline, and Amethyst. The interior of our lung surface is constantly being barraged by particulates from our environment that we breathe in. Black Tourmaline will 'scrape' foreign substances off the surface of the lung's lining. Also on the surface are microscopic hairs called 'cilia' that act as a filtration system in the lung. The influence of black Tourmaline is so strong that, alone, it can cause danger to those cilia. As with green Tourmaline, we add pink or watermelon Tourmaline to protect the

179

integrity of the cilia. In the case of the lungs being the 'where' you will select the combination for the applicable specifics. For fluid in the lungs you would add the stone Fluorite. For cancer in the lungs it would be white Calcite, and so on. Black Tourmaline is also called for when one has breathed toxic gas, fumes, and other irritant materials.

Green Tourmaline (See also in "Endocrine System")

Magnesium aluminum borate makes up green Tourmaline. It is for blood glucose imbalances. Magnesium influences insulin production and function. Whether treating diabetes or hypoglycemia, green Tourmaline will be a part of the stone equation. Muscle fibers take up glucose (sugars) and either use it immediately, or store it in the form of glycogen to be later broken down into glucose. Green Tourmaline indicates the endocrine function of the pancreas that regulates insulin production.

Pink Tourmaline

Pink Tourmaline is a borosilicate of aluminum and alkali, with iron and magnesium. Pink, or watermelon Tourmaline, is 'gatekeeper' to the tourmaline tools. Crystals that are both green and pink are watermelon tourmaline, and are used as pink. The presence of lithium contributes pink or red coloration to both solid and mixed pinks. Because the action of Tourmalines is so strong, there must be a 'stop-gap' influence to prevent opposite polarization, in the case of the green for glucose level, and the black, for protecting the interior of the lung surface. We NEVER use a green or black Tourmaline without a pink one present in the combination as well.

Wood Opal

Opalized wood is a silica-based composite that replaces fossilized wood. Silica is known to benefit the ears in multiple ways. In Rock-Medicine it is used to indicate the locale of the ears. When combined with other stones we are able to further stipulate whether the action is against an infection, such as with Iolite, or fluids with Fluorite, and numerous other applications.

Ulexite

Ulexite is hydrated sodium calcium borate hydroxide. Boron helps regulate the body's sex hormones and norepinephrine in the brain, heightening sexual sensitivity. Boron is known to influence the male reproductive system. Alone with the Seven Cleansers, Ulexite treats men or women for infidelity. Infidelity is not limited to the disloyalty of one partner to another but encompasses our entire family unit. A person is unfaithful to their child, to their parent, to their sibling, or their housemate whenever they close the door to healthy communication. Fidelity is the candid and honest exchange of sharing ones thoughts with one another.

Combined with Citrine and Bornite, Ulexite completes the combination for male hormone and/or indicates targeting the pituitary gland.

Wulfenite

Wulfenite indicates the head or skull. Wulfenite is a molybdate mineral found in the oxidized zones of hydrothermal lead deposits. Wulfenite is an enigma in terms of its symmetry. The release of oxygen by early life was important in removing molybdenum from minerals into a soluble form in the early oceans, where it was used as a catalyst by single-celled organisms. The skull has 22 different bones in it, made up of 8 cranial and 14 facial. These bones protect the brain and our organs of vision, taste, hearing, equilibrium, and smell. They anchor muscles that make up our facial movements and expressions.

Used with the Seven Cleansers and Amethyst, Wulfenite would be applied for any repair to a break, crack, fracture or surgical entry to any one of the 22 bones.

Wulfenite can be combined with Rhodonite for a facial nerve problem. It could be put with Rhodocrosite to say 'skull skin' which might be the scalp. With Halite it may be a facial muscle involved, although not being the jaw muscle, which is Halite and Aquamarine.

The Endocrine System

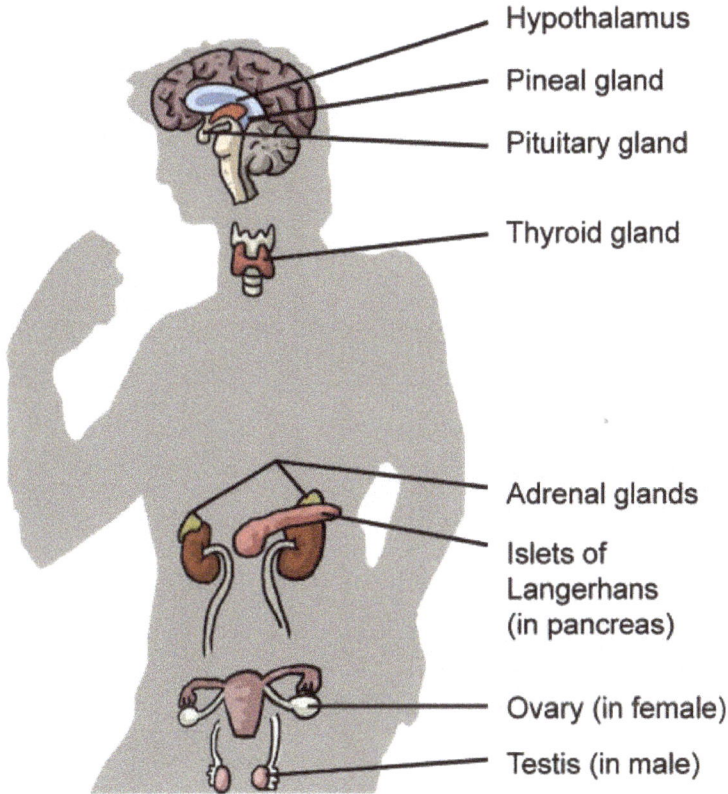

The Endocrine system is a complex and varied collective of glands and their support systems. There are two processes, the Endocrine and the Exocrine, represented by form and function.

The glands that secrete their products into ducts and out to the external environment are **exocrine**. Exocrine glands produce chemicals and some hormones. Most exocrine chemicals are secreted into ducts that lead directly to the external environment.

Although exocrine glands can produce a variety of chemicals, including hormones, they only deliver those chemical messages that travel and are received automatically, or 'unaware'. Typical exocrine glands include sweat glands, salivary glands, mammary glands, stomach, liver, and pancreas. They pass the remaining micro-fluids on to the endocrine to be secreted into the bloodstream. They pass their nanofluids, which include hormones, on to the endocrine delivery services. The nanofluids and hormones can carry physiological messages and emotional messages. They 'navigate' as well.

The glands that secrete their hormones directly into the bloodstream are **endocrine**. All other chemicals and hormones are carried to the endocrine system for delivery by secretion to the bloodstream. Hormones are chemical communication throughout the body. They influence virtually every cell and function of our anatomy as well as sexual development. They are responsible for the connection between the body and the emotions. Not unlike the principles behind the electrochemical communication between minerals, hormones contain *receptors* and *target cells*. This means they must have a chemical electric resonance that makes *sense* to an interaction before they will interact.

There are ten main parts to the endocrine system. The 8 glands are the pineal, pituitary, thyroid, thymus, adrenal, pancreas, ovary and testis.

Some hormones dictate organ function and are from **exocrine** glands. They use a duct system to move about the body. Some impact mood and emotion. Others regulate reproductive processes, including social responses and attraction or aversion between members of the same species. These are the **endocrine** hormones and are responsible for the olfactory phenomenon of human pheromones. Although more relevant to the psyche, healthy pheromones are the result of a healthy endocrine system and warrant a preview.

The endocrine system is responsible for not just what separates the men from the boys, but also what separates the men from the women. Those differences are both anatomical and emotional. Males and females think, and function, every bit as differently as they look. Hormone systems are what make females different from males...literally. Responsible for everything from breasts to babies, a female's hormone system drives not just her human existence but procreation itself. It is the woman's 28 day menstrual cycle that moves the process of reproduction. As her body goes through physical changes, a woman's hormone system is secreting chemicals. Some of these hormones, like oxytocin, send off signals that attract and stimulate sexual inspiration, or animal magnetism. Animal magnetism is driven by the primordial *will* to survive. Every element and organism shares a connection through this life force.

We began as simple organisms and have evolved into the amazingly intricate individuals that we are. When *'will'* evolved into *'desire'* the human being was born. We are creators and have just as much capacity for good as for ill. For too long we had medicine that was science devoid of Spirit. Popular metaphysics countered with medicine that was Spirit devoid of science. This is all too often seen among crystal users who assign no more specific use to rocks, gems and stones than that of their own intent. Desire can be dangerously creative, and even more obscure as allure. The will to exist is universal to all life forms, including pathogens such as bacteria. Pathogens are defined as biological agents that cause disease to its host. They live by replication, not reproduction. They have no inherent instinct because they are a toxic agent, and as such, an anti-life form. Although many organic life forms, including plants, communicate intent through hormones and pheromones, pathogens do not. Disease has no means of cooperation. Cooperation requires both communication and intent, giving the natural world a great advantage over disease.

Illness has no source for reinforcement, where human beings have remarkably cooperative systems and vast biological resources. The desire for wellness is called *vigor*, or the fluid life. Vigor is to infirmity as light is to darkness. No amount of darkness can put out a single light. The endocrine system is the fluid of communication for the human body's set point, or balance. The more advanced the endocrine system, the better equipped for survival as the fittest. The complexity of the form and function of the endocrine system is matched by the diversity of Rock-Medicine. There is a different stone formula to designate each one of the nine parts to the endocrine system. The combinations represent the 'form' of the gland. Other stones will be added to describe the receptor being sought. It may be muscular, vascular, neurological or otherwise.

Selenite - Parathyroid

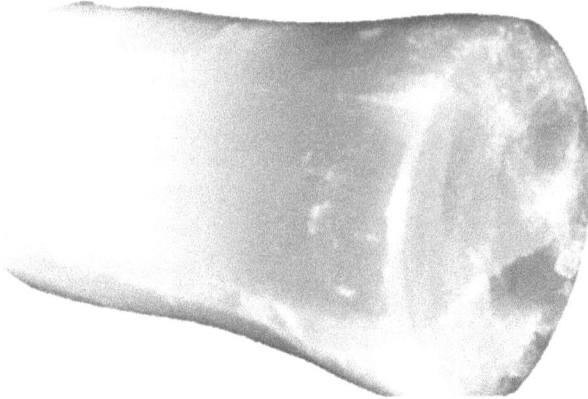

Selenite is for the parathyroid. It is a form of gypsum (not synonymous). It is the chemical calcium sulfate. The parathyroid controls the calcium levels in the blood stream, which in turn regulate the release of the parathyroid hormone. Inorganic calcium sulfate is an endocrine disruptor. Research shows a significant relevance between calcium sulfate and the reproductive systems of both genders. Puberty is the biochemical bridge from childhood to adulthood. It involves radical changes in the biological, psychological, cognitive and social processes.

Establishing the moment of puberty is by no means an exact science. However there are considerable physical changes in both male and female that can be observed as indicators. In girls, puberty typically occurs as early as age nine and as late as fourteen.

Male puberty ranges between ages ten and seventeen.

Every human body produces both the male sex hormone testosterone, and the female sex hormones estrogen/progesterone. It is their amounts that differentiate between the genders. Women use both the estrogen/progesterone and the small amount of testosterone their bodies produce. Males produce both progesterone, which gets converted to testosterone, and estrogen that they do not use. In fact, elevated estrogen levels in men presents health risks.

It has been observed that when sex hormone production commences in girls, there is a correlation with both thyroid activity and increased selenium levels. Physical manifestations of puberty include menstruation, breast development and increased body hair.

Male puberty is when the brain tells the pituitary gland to signal the testes to produce testosterone. The rapid increase in testosterone levels is coupled with dramatic responses involving, again, the thyroid and also selenium levels. Deepening of the voice, accelerated muscle development and body hair growth are some of the physical characteristics of puberty.

So, aside from treating the parathyroid, Selenite can be used to indicate that whatever damage the stone sentence is identifying occurred *before* puberty, and allows us to treat post-traumatically. In communicating a time reference, Selenite lets us pinpoint an occurrence of imbalance relative to any time prior to the biochemical processes that define puberty. We can use it to treat rape trauma that occurred when one was a child or a head injury that happened in childhood. Any adult addressing an issue of physical, emotional or spiritual trauma that occurred before puberty will include Selenite.

Apatite - Hypothalamus

Apatite is a phosphate, which designates the actual physical hypothalamus. The hypothalamus is considered a part of the Autonomic Nervous System *and* part of the Endocrine System. Because the hypothalamus has these two distinct communication references, it has multifaceted mineral combinations. If the hypothalamus is being treated in its functional capacity as part of the ANS, Rhodonite will be added to indicate that neurological association. (See "Rhodonite")

Although the hypothalamus is integral to the function of the endocrine system, it is not itself one of the glands or technical parts of the endocrine system. Managing both the body's metabolism and its hormone regulation, the hypothalamus has the job of homeostasis or "maintaining normal." The hypothalamus could be considered the physical manifestation of *vigor*. Sleep, thirst, blood pressure, temperature, reproduction, and electrolyte balances all fall under the jurisdiction of the hypothalamus. It is the neurotransmitter conduit between the brain and the pituitary gland for all hormone production and regulation. This link is called the **hypothalamic-pituitary-adrenal axis**. This is where the body meets the mind.

If the hypothalamus has structural damage or malady, the application would be a base of Seven Cleansers, Amethyst and Apatite. Absent any other stones, that combination, if applied, would do general cleansing of the hypothalamus.

If you add Garnet, which indicates circulatory system, you have now specified the **hypophyseal portal system.** If you exchange the Garnet for a Rhodocrosite, which is for epithelials, you have just targeted the Thymic **epithelial cells**. If you put the Garnet back in, you have reintroduced the circulatory system. However since the Rhodocrosite is still there, the total stone sentence is now describing epithelials of the circulatory system located in the hypothalamus, and that is the **endothelium** of the hypothalamus. Obviously, for novices this is a good example of when a precise diagnosis is invaluable, and also an understanding of the physiology of the diagnosis.

Ruby – Thalamus

Ruby is for the thalamus. The thalamus has generally been regarded as a group of relay nuclei that served as a 'gate' for sexual information from the spinal cord towards higher centers. Ruby, aluminum oxide, increases testosterone and estrogen-related gene expression as it relates to libido and pheromone activity. This pathway is thought to transmit peripheral sexual sensations.

It is the coming together of these sexual complexities that reissues life force and all that it is. We begin as energy and develop within the parameters of our dimension. We are in the third dimensional, or physical presentation, at present. The life force has been reinventing itself, with every new life and all that it holds, for a very long time. This is evolution. Ruby indicates the thalamus in the physical body. In the emotional body it would send the message that the treatment has a sexual consideration. One example might be to combine it with emotional stones when treating a sexual trauma or frigidity.

Hormone Balance Triad – Pituitary Gland

The pituitary gland is referred to as the *master gland* in the endocrine system. The pituitary gland is the communication relay between the hypothalamus in the brain and rest of the endocrine system. The hypothalamus communicates with the pituitary gland through the **posterior pituitary** and the **anterior pituitary**.

These six stones each have an individual use and application as well as combining with their respective counterparts for identifying the pituitary gland or its functions:

Male Pituitary	Female Pituitary
Bornite + Citrine + Ulexite	Lapis + Opal + Pearl

Some secretions, functions and forms of the endocrine system are different for a female than for a male. There is a combination of three stones for female anterior pituitary. And there are three stones that combine for male anterior pituitary. Beginning with the Seven Cleansers with Amethyst you will choose either three stone combination that together indicate male anterior pituitary **or** female anterior pituitary. If Lepidolite is added, indicating brain, the sentence switches the focus to a "part of the pituitary gland located in the brain". That is an exact description of the posterior pituitary.

The posterior pituitary regulates exocrine secretions and the anterior regulates endocrine secretions. And with the Seven Cleansers and Amethyst added, you have a treatment for the pituitary gland that can be further specified with additional stones as they may apply.

Azurmalachite – Pineal

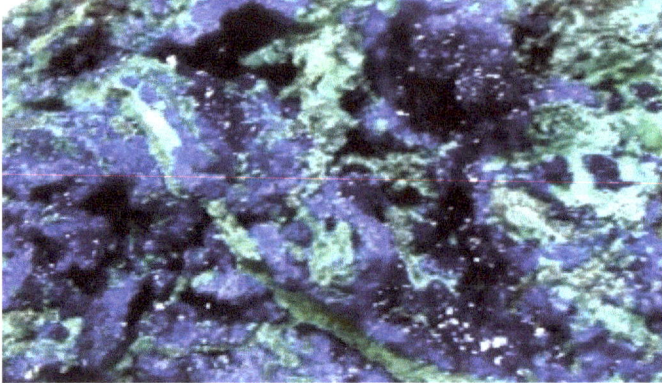

Azurmalachite is for the pineal gland. We use a single piece of material that contains both Azurite and malachite naturally. The pineal gland converts serotonin to melatonin. Serotonin is produced in the pineal gland, in the eye's retina, and in the GI tract. Its production is inhibited by light and promoted by darkness. Impacting our sleep patterns, melatonin has an impact on mood, metabolism and appetite. Only two percent of the total serotonin occurs in the brain. This, in turn, is distributed to multiple parts of the brain. Azurmalachite regulates the fraction that is utilized by the pineal gland for melatonin production.

As with the other exocrine glands in the endocrine system, the pineal is non-gender specific, so if you have the Seven Cleansers with Amethyst and Azurmalachite, you are treating the physical integrity of the pineal gland. When you add the stones for male or female balance you make it gender specific, which speaks to function and not form. It is the function of the pineal, and most other glands, that differs from male to female, not the form.

The melatonin hormones produced by the pineal gland are directed to the thyroid. When melatonin levels are disrupted depression can result. As the chemical imbalance progresses schizophrenia is the end result. This is basically 'fear as an illness'. Schizophrenia is now in the top ten disabling conditions affecting human beings. There are no laboratory tests available that are used for definite diagnosis. Rather there are symptoms ranging from depression, apathy and the management of emotions and thought processes to hallucinations, delusions, and hearing voices. Not considered curable, schizophrenia is treatable, but the side effects are such that many sufferers often discontinue their medications. Where the thyroid gland is the seat of inspiration, the pineal gland is the seat of imagination. So, in the pineal gland the creative visualization occurs, while in the thyroid the motivation follows.

Bloodstone (Heliotrope) – Thymus

Bloodstone is for the thymus, and is instrumental in the treatment of all viruses. The thymus is an important part of the immune system. It tells the body which substances belong in the body and which does not. Part of the endocrine system, the thymus has only one function: to produce t-cells. Killer T-cells are a subgroup of T-cells that kill cells that are infected with viruses. Viruses cause a wide range of infections and disorders like chicken pox, herpes, viral pneumonia, influenza and HIV. Bloodstone, or Heliotrope, is a compound of iron oxide. Iron Oxide induces human micro vascular endothelial uptake necessary for the production of human immune

T-cells. Without proper count levels, the immune system is weakened, thus allowing repeated infections and other health problems to occur more easily.

Being for t-cell production, Bloodstone has significant influence in the treatment of HIV and AIDS. Having the Human Immunodeficiency Virus does not mean one has AIDS. Acquired Immune Deficiency Syndrome is the result of having the HIV virus. Two primary things separate a patient from the status of being HIV positive and having their diagnosis be AIDS. One is that they must be HIV positive and the other is that their CD4, or t-cell count is below 200.

The attack on the immune system and resultant breakdown of natural defenses makes one vulnerable to virtually all-invasive infections. Bloodstone is specifically for t-cell production aiding in the elimination of the HIV presence from the bloodstream resulting in an HIV negative status. In the case of HIV we follow the recommended general cleansing with the specific combination of Seven Cleansers, Amethyst, and Bloodstone. The HIV virus levels present in the blood will steadily drop ultimately to a point of being undetectable. This will take a treatment schedule based upon the age of the patient, severity of their condition and longevity of their condition. When someone has AIDS, additional individual combinations may be given for lesions, brain infections, lung conditions, etc.

Of course, as with all illness, it is most beneficial to treat the afflicted with the utmost kindness and regard for their dignity. Nowhere more than in dealing with this deadly disease has this issue come forth. We must always remember this lesson in caring for that, which is ill of body, mind or spirit.

Seven Cleansers with Amethyst are recommended for all health workers and care givers to use as preventative measure when working with any communicable viral disease. Seven Cleansers used preventatively will boost the immune system's protection to its intended state of efficiency.

Rock-Medicine is so safe and inexpensive it is the only practical treatment to date for the epic HIV outbreaks in third world populations. HIV and other rampant contagions can be eradicated and prevented by simple cost effective methods as easily obtained as salinated distilled water.

Agate – Adrenal

Agate is for the adrenal gland and is comprised of silicon dioxide, which supports adrenal function. Typically agates form in ancient volcanic rocks or lavas where they represent cavities originally produced by the disengagement of elemental chemicals, called volatiles. The adrenal gland consists of the medulla and the cortex.

The **medulla** manufactures epinephrine and norepinephrine or, adrenaline. Adrenaline production is triggered by high-stress or physically exhilarating situations. The term, "fight or flight" came from the physical and emotional response to an increase in adrenaline levels. Using Agate alone is a way to give immediate response to sudden emotional trauma and the elevated adrenaline levels that come with it. Agate with the Seven Cleansers and Amethyst would be the base treating the form or integrity of the medulla.

The **cortex** is the where the three types of steroid hormones are produced. Steroid hormones are secreted by the gonads, or testes, adrenal cortex, and placenta. With a broad spectrum of distribution for body processes, these hormones affect everything from electrolytes to stress response. Having significant impact on many sexual developments, you will include the gender specific triad of hormone stones.

Yellow Calcite- Pancreas

Yellow calcite is the most stable of the forms of polymorphed calcium carbonate. Pancreatic juice is supersaturated with respect to calcium carbonate. The pancreas is both an endocrine gland producing several important hormones, including insulin, glucagon and somatostatin, as well as an exocrine gland, secreting pancreatic juice containing digestive enzymes. These enzymes help in the further breakdown of the carbohydrates, protein, and fat in partially digested materials that pass to the small intestine.

Cancer and diabetes are the two diseases most often linked to the pancreas. Pancreatic cancer is malignant neoplasm, and is addressed by the organ influence of the stone yellow Calcite, representing the form of the pancreas. We may from there add stones that pinpoint any vessel, epithelial, muscle, nerve or void that is infirmed. When influencing the pancreas for its endocrinal function, Rock-Medicine mediates cell-to-cell communication through local increases in the concentration of extracellular Ca^{2+}, co-released with insulin. Treating diabetes also involves two specific types, Type 1 and Type 2. Type 1 is genetic and Type 2 is based on consumption behavior. Those with Type 1 diabetes and the majority of advanced Type 2 currently require insulin injections. In those cases where insulin is currently required we add the stone Diopside, which is a pyroxene mineral.

These are exocrine processes and represent the 'glandular function of the pancreas which is only half its job. The pancreas also functions as an organ.

To distinguish between which function is being addressed, we include or exclude green Tourmaline with a pink Tourmaline, which is lithium based and affects islet insulin release. (See "Green Tourmaline")

In the absence of green Tourmaline, the stone combination will communicate with the pancreas for its organ function. This references the pancreas as an *exocrine* gland producing juices and enzymes sent to the small intestine to aid in digestion.

In Conclusion

Rock-Medicine is a resurgence of ancient knowledge that has evolved to meet modern day needs. It has taken over thirty years to develop the information as is contained in this book. It will take additional clinical research and double-blind studies to validate its content. Until then we have the ability to initiate use of the system. This will enable us to begin the wellness process as we gather case studies and testimonials to call attention to the need for investigation into its credibility.

The only hope for the future of humankind is that wellness is embraced and perpetuated. Whether you treat only yourself or take on the charge to set up blanket spreads that treat the whole planet, please participate. It is crucial that we establish standards with regard to the utilization of rocks, stones and gems, so they may be incorporated into our general healthcare system. There is no "guts or glory" in perpetuating an ideology of supernatural or imagined uses of these instruments. To continue to do so is, at its core, deceptive.

As you proceed with Rock-Medicine be mindful of how important it can be to inform your physician. Rock-Medicine is powerful healing and will be noticed by them. Have the courage to be honest for the sake of us all.

The unknown awaits only discovery to become the known.

Be well, Sela

About the Author - Sela Weidemann

In many languages ...
Selas birth name literally translates to 'rock shepherd'. Sela was born and raised in the Philadelphia area where her father worked as a chemist and patent attorney. In her early twenties she attended a lecture on the Hypothesis of Healing with Gems and Crystals; which inspired her to seek the truth in science to confirm the mineral kingdom as tools for wellness. The original information was gleaned one stone at a time over several years; through various methods; using kinesiology and known chemical standards. She compiled the basic information and wrote the first book, Rock-Medicine, Earth's Healing Stones From A-Z. After 25 years of consistent results, she has refined that knowledge to take our understanding of this system to the next level. With 5 years of research into biochemistry and verification via cutting edge medical advances; she offers this book for those who wish to use rocks, stones, crystals and gems as she does...for medicine. Sela now resides in small town in Arkansas, the crystal state with her dear husband Paul.

Glossary

- Alchemy- a form of chemistry and speculative philosophy

- Aluminosilicates- any naturally occurring aluminum silicate containing alkali-metal or alkaline-earth-metal ions

- Application- one individual treatment with any one of four available methods of use

- Bacteria – one celled organisms that can cause disease.

- Bio-chemical- the chemistry of living matter.

- Blanket spread- using quartz to send a stone or stones' chemical vibration a vast distance in all possible directions

- Charm Quark- a particular subatomic particle defined in Quantum Physics

- Combinations- more than one stone in a grouping that equals the treatment of a specific ailment

- Dark Matter- a form of matter invisible to electromagnetic radiation, which accounts for gravitational forces observed in the universe.

- Electromagnetic- of or relating to electromagnetism or electromagnetic fields.

- Electrons- an elementary particle that is a fundamental constituent of matter, having a negative charge

- Essence- rendering a stone or stones' chemical vibration into a salt solution to take as a remedy

- Ether- a substance that occupies all space, postulated to account for the propagation of electromagnetic radiation through space.
- Fermi Surface- abstract boundary in reciprocal space useful for predicting the thermal, electrical, magnetic, and optical properties of metals, and semimetals.
- Focus direct- using a quartz crystal with perfect point(s) to send a stone or stones' chemical vibration a distance in a straight line
- Fungus: A single-celled or multicellular organism. Fungi can be pathogens (such as histoplasmosis and coccidioidomycosis) that cause infections.
- Hand held- holding a stone or stones on one's hand for up to, but not exceeding, 20 minutes
- Holistic- inclusive of mind, body and spirit
- Homogeneous- of the same kind or nature; essentially alike.
- Immune System- a diffuse, complex network of interacting cells, cell products, and cell-forming tissues that protects the body from pathogens and other foreign substances
- Irradiating- to expose to radiation.
- Matrix- the material from which a mineral originates, takes form, or develops

- Neutrinos- any of the massless or nearly massless electrically neutral leptons.
- Neutrons- an elementary particle having no charge, mass slightly greater than that of a proton, and spin of $\frac{1}{2}$: a constituent of the nuclei of all atoms except those of hydrogen.
- Paramagnetism- a body or substance that, placed in a magnetic field, possesses magnetization in direct proportion to the field strength; a substance in which the magnetic moments of the atoms are not aligned.
- Paramagnetic Inertia- measure the interactions between the magnetic moments of the nuclei or electrons and the magnetic fields applied to them
- Parasitic Protozoa - A parasite is an organism that lives on or in a host organism
- Pathogenic- capable of producing disease
- Peroxidase- any of a class of oxidoreductase enzymes that catalyze the oxidation of a compound by the decomposition of an organic peroxide.
- PH level- used to express the acidity or alkalinity
- Positrons- an elementary particle having the same mass and spin as an electron but having a positive charge equal in magnitude to that of the electron's negative charge; the antiparticle of the electron.

- Protons- a positively charged elementary particle that is a fundamental constituent of all atomic nuclei
- Quantum- the smallest quantity of radiant energy.
- Quark- any type of subatomic particle with spin
- Resonance- the vibration produced
- Seven Cleansers-
- Jade – To remove the blockage of the flow of wellness.
- Amber – To link to the DNA 'memory' of your well state.
- Hematite – To purify the blood in people under 55 years old.
- Carnelian – To purify the blood in people 55 years and older.
- Smokey Quartz – To purify the water in the body.
- Pyrite – To purify the oxygen in the body.
- Clay – To regulate the immune systems.
- Cobalt – To remove radiation toxicity from the thyroid.
- Shaman- a person of Asiatic origin, who acts as intermediary between the natural and supernatural worlds to cure illness
- Sub atomic- noting or pertaining to a particle or particles contained in an atom, as electrons, protons, or neutrons.
- Thyroid Peroxidase- (TPO) is an enzyme expressed mainly in the thyroid where it is secreted into colloid.
- Universal Mass- The sum of all matter

- Virus-an infective agent that typically consists of a nucleic acid molecule in a protein coat, is too small to be seen by light microscopy, and is able to multiply only within the living cells of a host.
- Water-soluble- capable of dissolving in water.

Notes

Notes

www.ingramcontent.com/pod-product-compliance
Lightning Source LLC
Chambersburg PA
CBHW041256040426
42334CB00028BA/3036